Finite Math on the Web... What's in it for you?

- **A better understanding of finite mathematics.**

- **More help with homework and tests.**

- **A better grade.**

This unique course companion combines the power of the Web, a CD-ROM, and a practical workbook to give you the tools you need to be successful in your course work. Ten modules, each with its own **interactive Java Applets**, help you explore, visualize, and understand even tough finite math concepts. These modules correspond exactly to the chapters in the student manual and to the modules on the Web site (www.FiniteMathTutor.com).

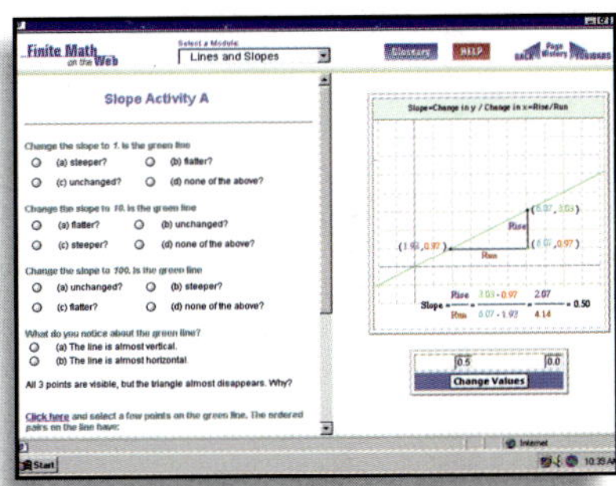

Module 1 Lines and Slopes — This module introduces the concepts of slope of a line, parallel lines, and perpendicular lines. Students learn how to compute the slope and intercepts of a line, as well as how to find the intersection of two non-parallel lines.

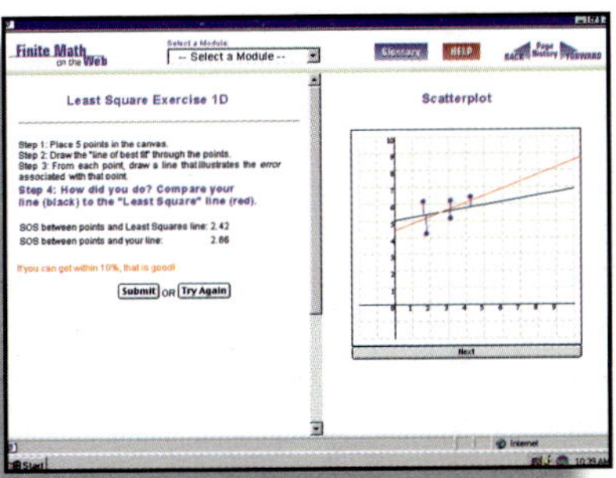

Module 2 Least Squares — This module introduces the concept of a "regression line" which fits data in the best possible manner. The dependence of the slope on the location and number of data points is shown through an applet.

Module 3 Matrices and Linear Systems — This module introduces the concept of a system of linear equations. Matrix notation is introduced along with the basic operations of addition, subtraction and multiplication. Matrix inverses are explained, and various ways of solving linear systems using matrices are discussed.

www.FiniteMathTutor.com

Finite Math on the Web... lets you do it all!

Module 9 **Probability Distributions** — This module introduces the binomial distribution, the normal distribution, and the concept of a random variable. It also develops the central limit theorem in the form of the "Law of Large Numbers."

Module 10 **Financial Applications** — This module introduces interest, both simple and compound, then the exponential function as the limit of continuous compounding. Annuities and a simple stock market simulation are introduced.

Each module lets you

boost your understanding by working interactive exercises and then check your progress with practice quizzes that give you an instant score!

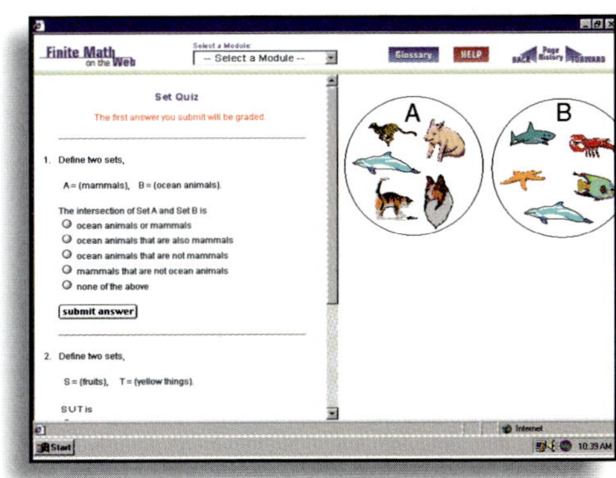

Finite Math on the Web... lets you do it all!

Get ready for exams by reviewing terms quickly in the online glossary!

A help guide and list of Frequently Asked Questions is included on the Web Site and on the CD-ROM, so you are never far from answers.

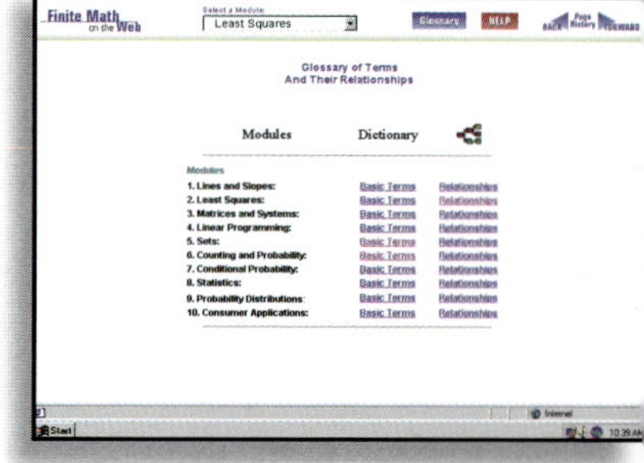

Once you've got the ins and outs of finite math down, follow the links and explore it more with online explorations!

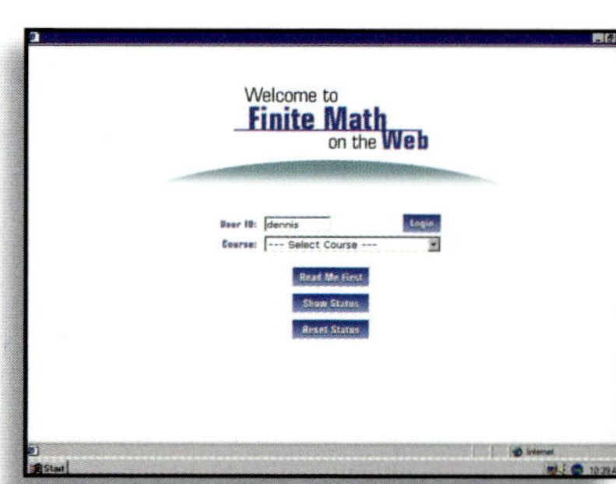

Want to launch your success in Finite Math?
Log on to **www.FiniteMathTutor.com**
to launch Finite Math on the Web!

DUXBURY

FINITE MATH ON THE WEB

Michael S. Pilant
Janice L. Epstein
Kathryn Bollinger
Robert Hall
Yvette Hester
Arlen Strader

Texas A&M University
College Station, Texas

THOMSON
™
BROOKS/COLE

Australia • Canada • Mexico • Singapore • Spain
United Kingdom • United States

Publisher: Curt Hinrichs
Assistant Editor Ann Day
Editorial Assistant: Katherine Brayton
Technology Project Manager: Earl Perry
Marketing Manager: Joseph Rogove
Marketing Assistant: Jessica Perry
Project Manager, Editorial Production: Janet Hill

Print/Media Buyer: Doreen Suruki
Cover Designer: Denise Davidson
Cover Image: Getty Images
Cover Printer: Webcom Limited
Printer: Webcom Limited

Printed in Canada

1   2   3   4   5   6   7   07   06   05   04   03

For more information about our products,
contact us at:
Thomson Learning Academic Resource Center
1-800-423-0563
For permission to use material from this text,
contact us by:
Phone: 1-800-730-2214
Fax: 1-800-730-2215
Web: http://www.thomsonrights.com

ISBN 0-534-99757-0

Brooks/Cole Thomson Learning
10 Davis Drive
Belmont, CA 94002
USA

Asia
Thomson Learning
5 Shenton Way #01-01
UIC Building
Singapore 068808

Australia/New Zealand
Thomson Learning
102 Dodds Street
Southbank, Victoria 3006
Australia

Canada
Nelson
1120 Birchmount Road
Toronto, Ontario M1K 5G4
Canada

Europe/Middle East/Africa
Thomson Learning
High Holborn House
50/51 Bedford Row
London WC1R 4LR
United Kingdom

Latin America
Thomson Learning
Seneca, 53
Colonia Polanco
11560 Mexico D.F.
Mexico

Spain/Portugal
Paraninfo
Calle/Magallanes, 25
28015 Madrid, Spain

Contents

APPENDIX A   APPLETS   185

APPENDIX B   GLOSSARY   209

Preface

Finite Math on the Web is a web-based companion to the Finite Mathematics course taken by undergraduate students at many colleges and universities. *Finite Math on the Web* is composed of a series of ten web modules and a student manual with CD-ROM that integrates the web modules into a textbook-based course. Because the standard Finite Mathematics course is more topical than sequential, we have focused on those essential mathematical concepts that form the basis for topics for math education in the social sciences, life sciences, and business. Within each of these topics we then identified essential areas that students have traditionally had difficulty mastering. Our aim is to present these key mathematical concepts through JAVA applets in a visual and interactive way that cannot be done through textbooks.

Objectives

Our primary goal in developing the product was to address several teaching challenges we faced at Texas A&M University. We teach at a large university with thousands of students taking Finite Mathematics each year. While our course remains large, the resources available to effectively teach our students have decreased. We no longer are able to offer ample graduate student assistance for both one-on-one tutoring and the grading of homework. We developed *Finite Math on the Web* to help us retain the quality of course instruction and to better assess and manage the needs of our students.

The pedagogical objectives of *Finite Math on the Web* are threefold:

- To present key mathematical concepts, those that students struggle with the most, in a visual and interactive way that cannot be done effectively with a textbook. Through the use of JAVA applets for visualization and simulation, we have found that our students gain a better understanding of concepts and hone their intuition.

- To help those students who fear or struggle with mathematics. There has been a tremendous amount of research done on where and why students have difficulty with mathematical concepts. We have combined our own experience with this research to identify and focus on those areas. *Finite Math on the Web* offers immediate feedback to students, to encourage them as they work through the concepts.
- To motivate students. It is no secret that many students taking this course wonder why they need to take it. We have tried to make the learning experience fun.

Teaching and Learning Styles

This product has been class tested with over 14,000 students over the last five years, and we have collected a large amount of usage data to help us understand how students use this product and what effect, if any, it has on improving their comprehension of the target concepts. Our data suggests that success rates do improve when using the module courseware, algorithmic quizzes and workbook. The data also suggests that individual students use these resources in different ways. Some students make a game of it by working through the courseware several times quickly. Others may only work through the courseware once, but do so carefully. While students' study habits vary, *Finite Math on the Web* allows us to identify where each student is having difficulty and to track the student's progress.

Organization

The ten modules are arranged into two main tracks. The first track is oriented to algebraic computations, geared towards graphing, solving linear systems of equations, fitting curves to data and linear programming. The second is oriented toward sets, probability, and statistics. The module on Financial Applications is self-contained. The order in which the modules are accessed is determined entirely by the instructor. Within each module, the activities are logically sequenced, but students may progress through the activities in any order that they wish.

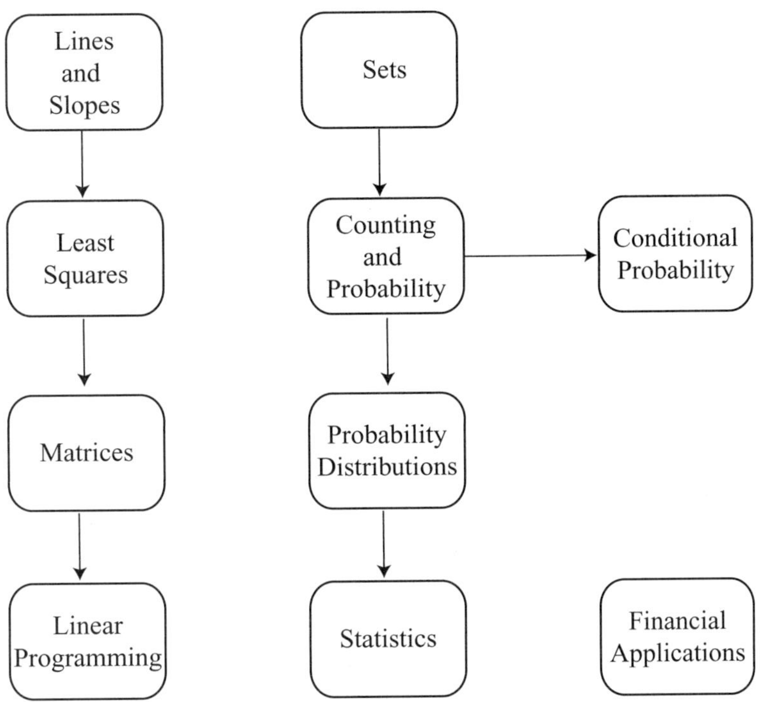

Each of the modules begins with an overview, a review (if necessary), and motivational material. Applets appear in simplified form to illustrate the basic concepts. Checkpoints, in the form of multiple-choice questions, give feedback to the student. As the concepts build in complexity, the student is allowed more interaction through the applets. Finally, at the end of each module there are ungraded exercises and a graded quiz. The exercises in the module may be repeated as often as desired, and the answers are provided. The results of the graded quiz are only available to the instructor.

The graded quizzes for the modules are administered through a separate courseware management product, *BCA*. The registration process, involving a unique PIN number, and enrollment in the correct course are outlined in the next section.

The web modules, CD, and student manual are intended to be used in roughly equal proportion. In addition, the web site at http://www.finitemathtutor.com/ contains numerous links to other resources for the student to use.

Requirements and Prerequisites

The prerequisites for *Finite Math on the Web* are the same as those for the classical text-based Finite Mathematics course. The only technology requirements are access to the Internet (with a 28.8K modem or better) and Internet Explorer 6.0. In order to use the additional resources found on the CD, access to a CD-ROM drive is required.

At the end of the student manual, there is a brief description of each applet, a Glossary, Answers to the Odd Exercises, and Answers to the Sample Quizzes. Full solutions to the odd exercises and sample quizzes in the manual are provided on the student CD.

The authors have found that the web modules provide a rich context for the interactive visualization of mathematical concepts so important to students' success in the life sciences, business, and the social sciences. We hope that students will also benefit from these materials.

Instructions for Registering

To fully utilize the features of *Finite Math on the Web* that allow an instructor to track student progress in the course, each student will need to register their copy of *Finite Math on the Web*. It is a simple process, illustrated as follows.

1. **Access the web site** — Begin by opening the page **bca.brookscole.com**. On your first visit to this site when you register, please choose the link "First Time Users." When you return to this site, please choose the "Login" link.

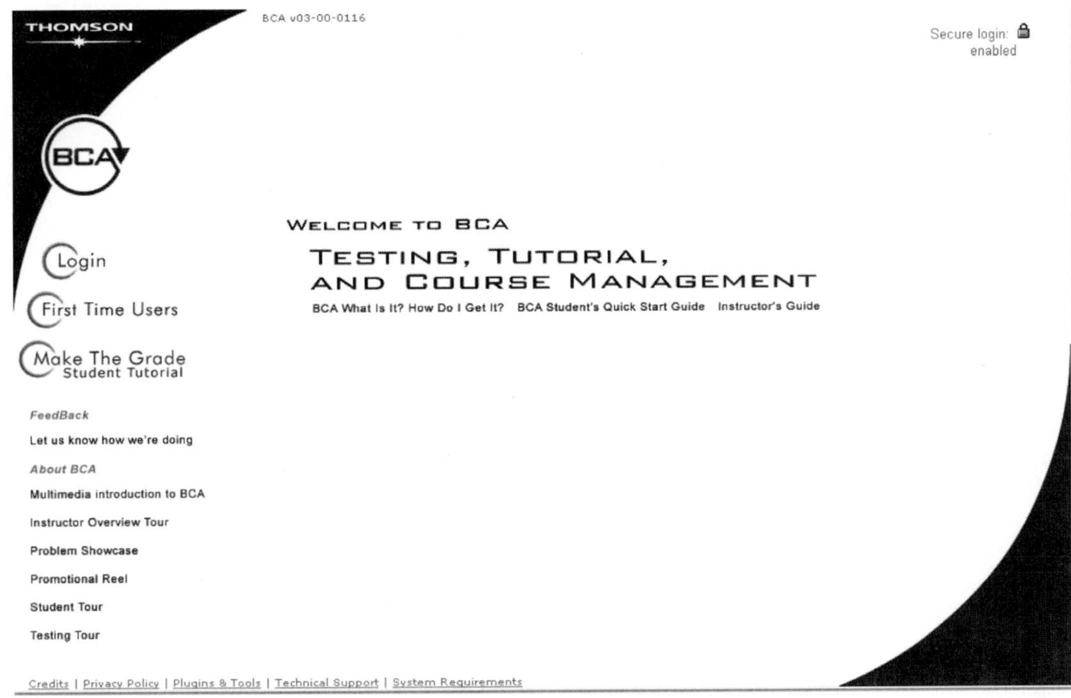

2. **Find your school** — The next screen you see will allow you to select your school. Start by selecting your *State* from the selection box and then type the first few letters of your school name into the *Name of school* box. In the example below, the state is Arizona and "u" appears in the search box. Select your school by clicking on the underlined link.

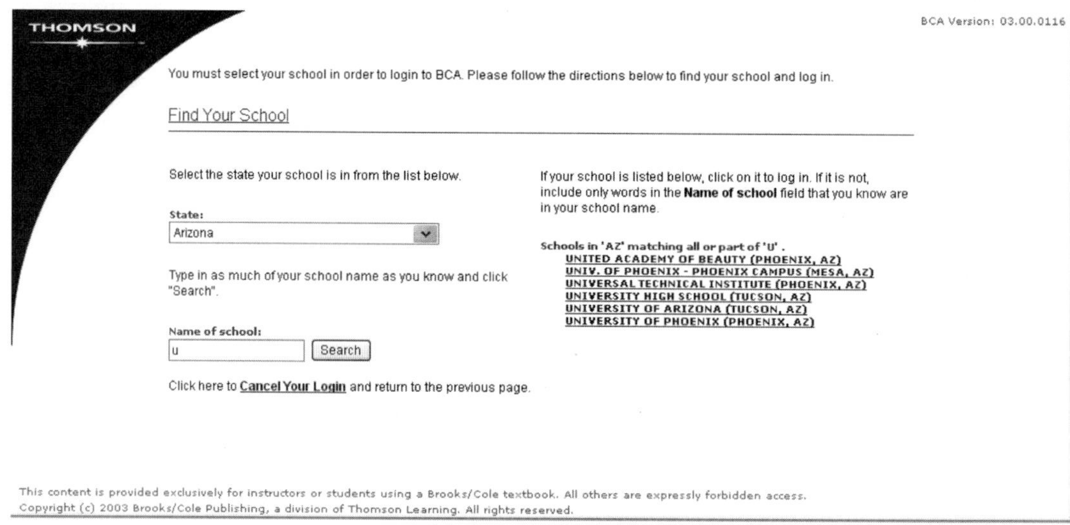

3. **Use your PIN code** — There is a CD pocket in every **NEW** copy of this manual. In that pocket, you will find a card containing a PIN code. After selecting your school, carefully enter this code into the PIN code box, as shown below.

    Note: *This PIN code is case-sensitive. If you encounter an error with your PIN code, please contact Tech Support. (There is a link to Tech Support in the upper right corner of this screen.)*

    If you have a used copy of this book and the PIN code has been used previously, you will need to purchase a new PIN code at **www.finitemathtutor.com**. Your PIN code is unique and can only be registered once. Protect your PIN code; don't let anyone else use it.

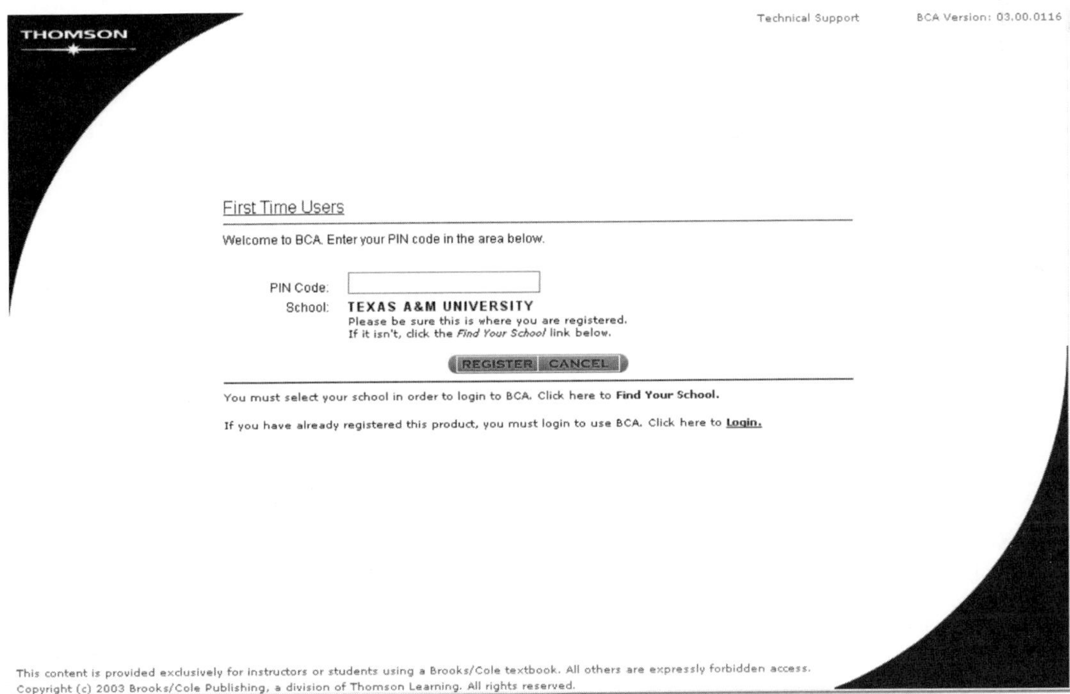

4. **Fill in your registration** — Simply fill out the registration form, being careful and accurate. Select your own User Name and Password carefully to protect the security and integrity of your records. If your name is John Smith, don't use "john" as your User Name and "smith" as your Password. Make choices that you will remember but others will not easily guess. BCA is a very secure system when used wisely.

Welcome to BCA Registration

You must be a student or an instructor using an authorized textbook to register. All others are expressly forbidden access.

You will only need to enter this information once per book or course.
If you have already registered, you need to enter your login name and password <u>here</u>. You will be registered for the desired material once you have logged in.

Before you register, please read our <u>Terms of Service</u> agreement which contains important information about our privacy policy, copyright information, etc. that you should understand before registering. By registering, you indicate that you have read and agreed to our Terms of Service.

You have chosen to register for the Book : Finite Math on the Web

Located at: TEXAS A&M UNIVERSITY

To complete your registration, please enter the following information. (* - indicates a required field)

1. Enter Your User Information

* First Name:
* Last Name:
* E-mail Address:
Department:
City:
State: (choose one)
Country:
Postal Code:

2. Choose Your Login Name And Password

Note: Login names and passwords are case sensitive.

* Login name:
2-30 characters, letters, numbers & '_'
* Password:
4-15 characters, letters, numbers & '_'
* Re-enter Password:
* Verification Question: We will ask this question if you forget your password.
* Verification Answer: This is the answer we will expect in response to the question above.

Once you have registered, you will be sent to your assignment page.

5. **Enroll in Your Course** — The next step in the registration process is to enroll in your class. This will allow you to take and view the assignments your instructor has created. Begin by choosing the link in the left column labeled Self Enrollment. If you do not see your course listed, choose the link at the bottom of the page to enter a Course Enrollment PIN. You will need to get a course PIN code from your instructor to complete this method of enrollment.

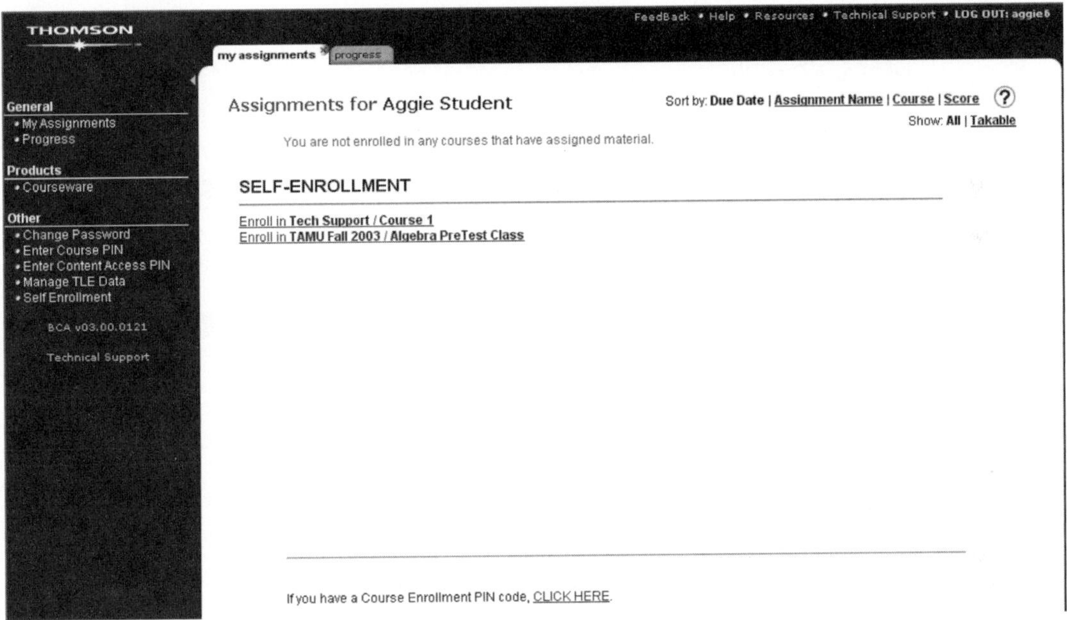

6. **Getting Started** — After enrolling in your class, you will be able to take the assignments that your instructor has assigned. Additionally, in the left column under Products there is a link to Courseware. To access the module contents, choose this Courseware link and then click on the image. These modules are designed to prepare you for the quizzes that are assigned. It is strongly suggested that you review the module associated with a quiz before attempting the quiz.

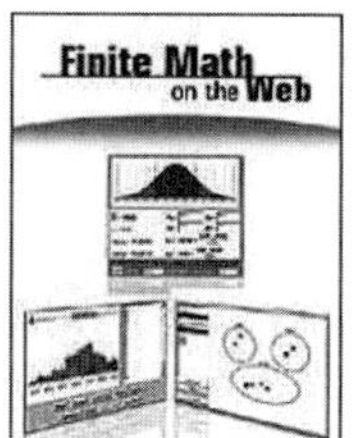

The individual modules are available in a pull-down menu found at the top of the main courseware screen shown below:

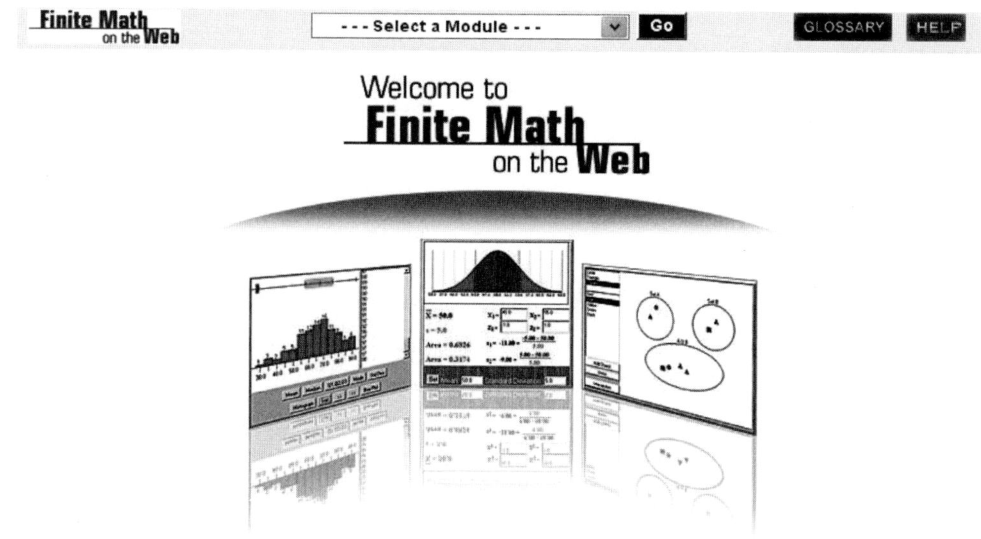

Acknowledgments

We would like to express our gratitude to all of the people whose efforts have contributed to this project, including the reviewers who made recommendations to the web modules at various stages of development.

James Balch
Middle Tennessee University

George Bergeman
Northern Virginia Community College

Jeanne Bowman
University of Cincinnati

Mary Davenport
Blinn College, Bryan Campus

Patrick Driscoll
U.S. Military Academy

Larry Ellerbauch
Northern Michigan University

Harvey Greenberg
University of Colorado at Denver

Dan Harned
Michigan State University

Katie Kolossa
Arizona State University

Frederick Lane
Palm Beach Community College

Raymond Lee
University of North Carolina, Pembroke

Mickey Levendusky
Pima Community College

Joseph McDonald
Community College of Southern Indiana

Thomas McFarland
University of Wisconsin, Whitewater

Yash Mittal
University of Arizona, Santa Rita

Elvira Munoz-Garcia
Arizona State University

Leela Rakesh
Central Michigan University

Anurag Singh
University of Utah

Mary Wagner-Krankel
St. Mary's University

We would also like to thank Art Belmonte, Bill Rundell, Al Boggess, Mila Mogilevsky and Tom Kiffe for their contributions to this project. A special thanks also goes the students at Texas A&M University, College Station, Texas.

Michael S. Pilant
Janice Epstein
Kathryn Bollinger
Robert Hall
Yvette Hester
Arlen Strader

Slopes and Lines

One of the most fundamental procedures in mathematics is to represent data points by means of a curve or a **graph**. For over two hundred years, graphing data points in two dimensions by means of a curve has proven to be one of the most effective ways to understand and visually represent numbers. The simplest relationship between data points is that of a straight line. In this module we review the basic properties of straight lines (including slopes and intercepts) and discuss several important linear models, including supply and demand.

1.1 REVIEW OF LINES

Intercepts are points that lie on the x- or y-axis. Intercepts are easy to find by setting to 0 the variable (axis) whose intercept you *do not* want (Table 1.1). The intercepts are shown in Figure 1.1.

name	location	ordered pair
y-intercept	$x = 0$	$(0, y_1)$
x-intercept	$y = 0$	$(x_1, 0)$

Table 1.1   Intercepts

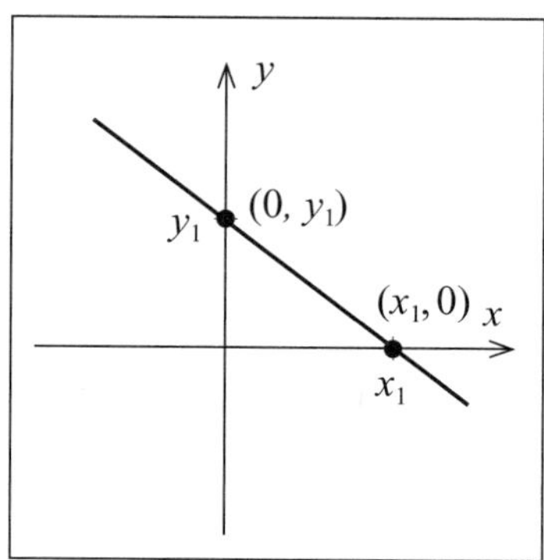

Figure 1.1 x- and y-Intercepts

Example 1.1   Find the x- and y-intercepts for the line $2x + 3y = 6$ and show them on a graph.

Solution

Set $x = 0$ to find the y-intercept:

$$2(0) + 3y = 6$$
$$3y = 6$$
$$y = 2$$

Thus, we say that the y-intercept is the ordered pair $(0, 2)$. Set $y = 0$ to find the x-intercept:

$$2x + 3(0) = 6$$
$$2x = 6$$
$$x = 3$$

Thus, we say that the x-intercept is the ordered pair $(3, 0)$. The intercepts for the line $2x + 3y = 6$ are shown in Figure 1.2.

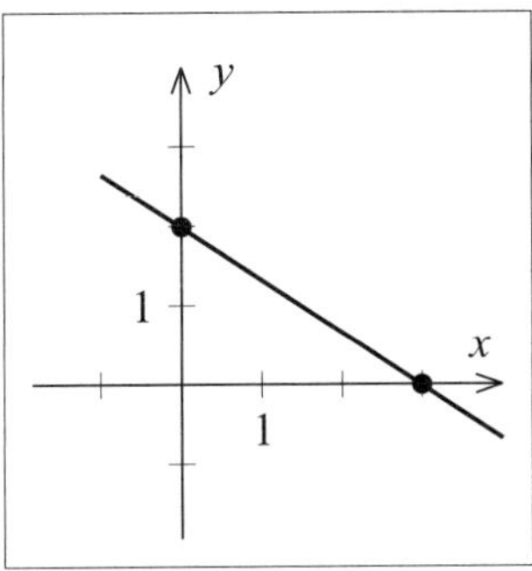

Figure 1.2　Graph of $2x + 3y = 6$ 

Slope is a measure of the rate of change of y with respect to x. Intuitively, slope is easily remembered as **rise** over **run** (where *rise* is the net change in height and *run* is the net change in length). But that doesn't help us find its **value**. We can also say that slope is the "change in y" over the "change in x," but that doesn't help us find its value either. We need to take the intuitive definitions of slope and find a formula that will give us a value for slope. To find the slope of a line, we need two points (two ordered pairs) on the line.

Let the first point be denoted (x_1, y_1) and the second point be denoted (x_2, y_2). Then, the slope is given by the *slope equation*

$$\text{slope} = m = \frac{rise}{run} = \frac{\Delta y}{\Delta x} = \frac{y_2 - y_1}{x_2 - x_1} = \frac{y_1 - y_2}{x_1 - x_2}$$

Note: *The slope* cannot *be given as the ratio*

$$\frac{y_2 - y_1}{x_1 - x_2}$$

You must be consistent once you pick which way you are going to subtract; you can't change your mind in the middle of things!

Example 1.2　Given the two points $(2, -1)$ and $(-3, 3)$, find the slope of the line between them.

Solution

We can calculate the slope this way:

$$\frac{3-(-1)}{-3-2} = \frac{4}{-5}$$

Or, we can calculate the slope this way:

$$\frac{-1-3}{2-(-3)} = \frac{-4}{5}$$

but *not* this way:

$$\frac{3-(-1)}{2-(-3)} = \frac{4}{5}$$

Thus, the slope is $-\frac{4}{5}$. This means that from anywhere on the line defined by these two points, we can get to another point on the line by moving *down 4 units* and to the *right 5 units*, or by moving *up 4 units* and to the *left 5 units*. When the two given points are graphed, it can be seen that the slope of the line that passes through these points must be negative (Figure 1.3a and 1.3b).

Note: *If the slope has the wrong sign, you know the points have been reversed.*

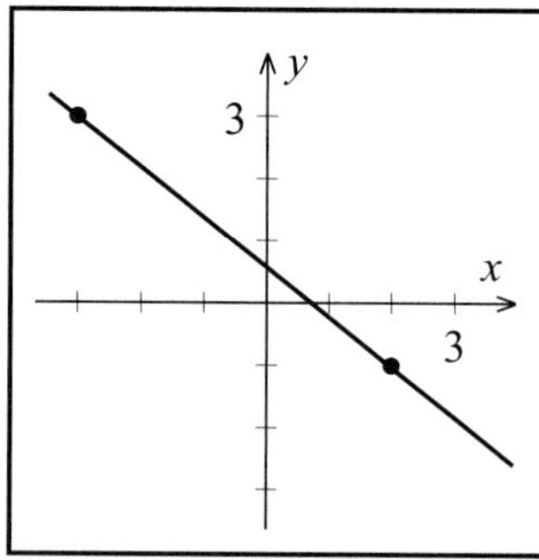

Figure 1.3a   Graph of Example 1.2 (Traditional Style)

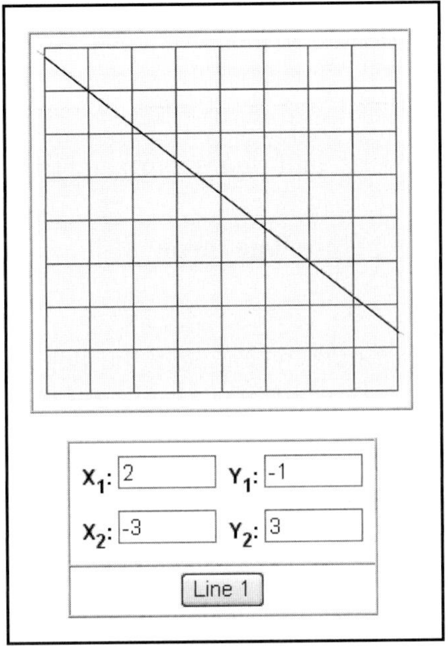

Figure 1.3b　Graph of Example 1.2 (Graphing Applet Style)

1.2　EQUATION OF A LINE

The **equation** of a line can be written in three different forms. The **general** or **standard form of a line** is given as

$$ax + by + c = 0$$

where a and b are *not both* 0. This form of a line is the easiest form in which to find the intercepts, as in Example 1.1.

The **slope-intercept form of a line** is given as

$$y = mx + b$$

This form of a line allows the slope m, and y-intercept b, to be read directly from the equation.

 **Example 1.3**   The line $y = 2x + 1$ is given in slope-intercept form. Find the slope and the y-intercept of this line.

Solution

The slope of this line is 2 (*up 2 units* and *right 1 unit*) and the y-intercept is at (0, 1).   ❖

The third form of a line, the **point-slope form,** is given as

$$y - y_1 = m(x - x_1)$$

This form of a line is really a formula; when given the slope and a point, we can find the equation of the line. Of course, we can also use this form if we are given just two points. From the two points we can find the slope, and then we can use this slope and our choice of one of the points to find an equation of the line.

 **Example 1.4**   Given the two points (1, −1) and (3, 4), find an equation of a line that passes through both points.

Solution

First, we must find the slope.

$$m = \frac{4 - (-1)}{3 - 1} = \frac{5}{2} = 2.5$$

Now, we pick one of the points to plug into the equation for (x_1, y_1).

Note: *Values are substituted into the equation only for the variables with subscripts.*

Use the point (1, −1). Substituting into the point-slope form, $y - y_1 = m(x - x_1)$:

$$y - (-1) = 2.5(x - 1)$$
$$y + 1 = 2.5x - 2.5$$
$$y = 2.5x - 3.5$$

The equation of the line is $y = 2.5x - 3.5$.   ❖

We summarize the behavior of lines in Table 1.2

Slope	Line	Equation	Example
$m > 0$	Rises	$y = mx + b$	$y = 2x + 1$
$m < 0$	Falls	$y = mx + b$	$y = -2x + 1$
$m = 0$	Horizontal	$y = b$	$y = 2$
m undefined	Vertical	$x = a$	$x = 1$

Table 1.2   Types of Lines

The last two equations in Table 1.2 introduce two special types of lines: horizontal and vertical lines. A **horizontal line** is a line that has a slope of 0. That means there is *no change in y*. Hence, the equation of the line is one where *y* is constant. A **vertical line** is a line that has an **undefined slope**. That means there is *no change in x*. Hence, the equation of the line is one where *x* is constant.

Example 1.5   Find an equation for each of the following lines:

a. Passes through the points $(1, -1)$ and $(2, -1)$.
b. Passes through the points $(1, -1)$ and $(1, 2)$.
c. Is horizontal and passes through the point $(1, -1)$.
d. Is vertical and passes through the point $(1, -1)$.

Solution

a. Notice that the *y*-coordinate of both points is the same, telling us that there is *no change in y*. Therefore, *y* is constant. What constant? There is only one constant *y* can be, -1. Hence, the equation of the line through both points is $y = -1$. Another method to find the equation of the line is to use the slope equation to find that

$$m = \frac{-1 - (-1)}{2 - 1} = \frac{0}{1} = 0$$

Since $m = 0$, we see from the summary chart that we have a horizontal line with equation $y = -1$. The Graphing Applet can show how the line will look when given these two points. The lower, darker line in Figure 1.4 is the horizontal line $y = -1$.

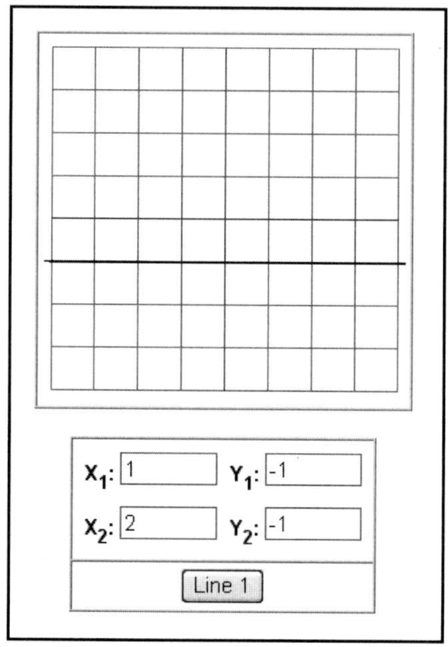

Figure 1.4   Applet Graph of $y = -1$

b. Notice that the x-coordinate of both points is the same, telling us that there is *no change in x*. Therefore, x is constant. What constant? There is only one constant x can be, 1. Hence, the equation of the line through both points is $x = 1$. An alternate method to find the equation starts with using the slope equation to find

$$m = \frac{2 - (-1)}{1 - 1} = \frac{3}{0} = \text{undefined}$$

Since m is undefined, we see from the summary chart that we have a vertical line with equation $x = 1$. The Graphing Applet can show how the line will look when given these two points. The rightmost, darker line in Figure 1.5 is the vertical line $x = 1$.

Note: *When using the graphing applet, you cannot enter vertical lines using a slope and an intercept since the slope is undefined. You MUST use two points to graph a vertical line.*

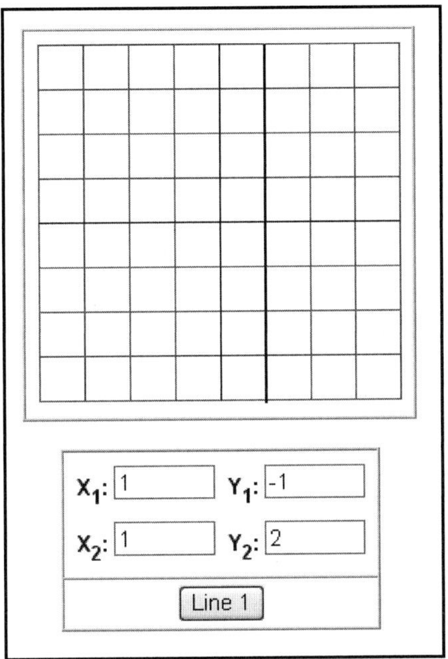

Figure 1.5   Applet Graph of $x = 1$

c. The only horizontal line that can pass through the point $(1, -1)$ is one that has the same y-value as the point itself. Hence, the horizontal line must be $y = -1$.

d. The only vertical line that can pass through the point $(1, -1)$ is one that has the same x-value as the point itself. Hence, the vertical line must be $x = 1$.   ❖

Note: *You should now complete Activities A, B, and C in the Lines and Slopes Module.*

1.3   PARALLEL AND PERPENDICULAR LINES

Parallel and **perpendicular** describe a relationship between two or more lines.

> Two lines are parallel if and only if they have the same slope,
> $$L_1 \mathbin{/\mkern-5mu/} L_2 \ \text{ iff } \ m_1 = m_2$$

The two lines $y = 2x + 1$ and $y = 2x + 3$ are parallel because they both have a slope of 2. The first line intersects the y-axis at $(0, 1)$ and the second line intersects the y-axis at $(0, 3)$.

> Two lines are perpendicular if and only if their slopes are negative reciprocals,
> $$L_1 \perp L_2 \ \text{ iff } \ m_1 = -\frac{1}{m_2}$$

The two lines $y = 2x + 1$ and $y = -\frac{1}{2}x + 1$ are perpendicular because their slopes are negative reciprocals. An alternative way to consider perpendicular lines is to say the product of their slopes is -1: $m_1 \times m_2 = -1$.

Example 1.6   Given the line $2x + 3y = 6$,

a. Find an equation of a line parallel to the given line that passes through the point $(-1, 3)$.
b. Find an equation of a line perpendicular to the given line that passes through the point $(-1, 3)$.

Solution

First, we must find the slope of the given line by writing it in slope-intercept ($y = mx + b$) form.

$$2x + 3y = 6$$
$$3y = -2x + 6$$
$$y = -\frac{2}{3}x + 2$$

Thus, the slope of the given line is $-\frac{2}{3}$. Now we are ready to begin answering the questions.

a. Any line that is parallel to the given line must have the same slope of $-\frac{2}{3}$. To find an equation of a line through the given point with slope $-\frac{2}{3}$, we can use the point-slope formula.

$$y - 3 = -\frac{2}{3}\left[x - (-1)\right]$$
$$y - 3 = -\frac{2}{3}(x + 1)$$
$$y - 3 = -\frac{2}{3}x - \frac{2}{3}$$
$$y = -\frac{2}{3}x + \left(3 - \frac{2}{3}\right)$$

Therefore, $y = -\frac{2}{3}x + \frac{7}{3}$ is an equation of a line passing through the point $(-1, 3)$ and parallel to the given line. We can use the Graphing Applet to verify that the lines are parallel (Figure 1.6).

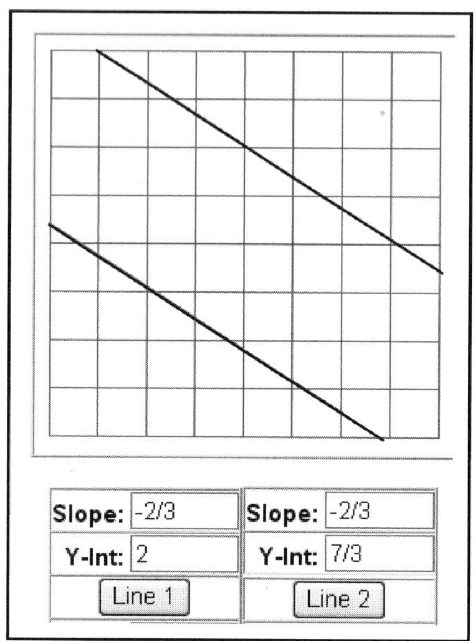

Figure 1.6   Applet Graph of Two Parallel Lines

b. Any line that is perpendicular to the given line must have a slope of $\frac{3}{2}$, the negative reciprocal of $-\frac{2}{3}$. To find an equation of a line through the given point with thegiven slope, we can go directly to the slope-intercept form.

$$3 = \frac{3}{2}(-1) + b$$

$$3 = -\frac{3}{2} + b$$

$$b = \frac{9}{2} = 4.5$$

Therefore, $y = 1.5x + 4.5$ is an equation of a line through $(-1, 3)$ and perpendicular to the given line. In Figure 1.7 we use the Graphing Applet to verify that the lines are perpendicular.

Note: *When graphing to check if lines are perpendicular, the scale on the x-axis and the y-axis must be the same for this visual verification to work.*

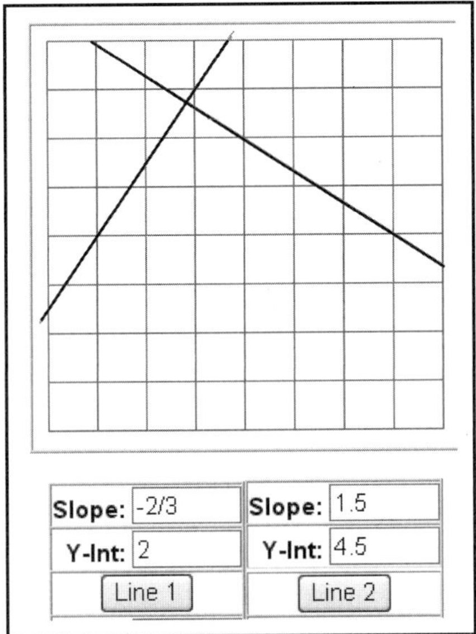

Figure 1.7 Applet Graph of Two Perpendicular Lines ❖

Either of the two methods above will work for finding an equation of a line, if both a point and the slope are known.

 **Note:** *You should now complete Activity D in the Lines and Slopes Module.*

1.4 MODELING WITH LINES

The previous review of lines was a necessary foundation on which to introduce and build the idea of **mathematical modeling** with lines. Mathematical modeling is the process of applying mathematics to solve real-world problems through mathematical descriptions of certain relationships. An important way of describing relationships is to use functions. A **function**, f, is a rule that assigns to each value of x *one and only one* value of y. The notation we use is

$$f(x) = y$$

where x is the **independent variable** and represents the **domain** (the set of all possible input values) and y is the **dependent variable** and represents the **range** (the set of all possible output values).

If the graph of $y = f(x)$ is a straight line, we know that $f(x)$ is a **linear function**. From the previous section, 1.2, we know that $y = mx + b$ or $f(x) = mx + b$ is a linear function. This representation can also be expressed as an ordered pair:

$$(x, y) = (x, f(x))$$

Example 1.7   Suppose you are given the function $f(x) = 2x - 3$. Find the ordered pair representation if $x = 1$.

Solution
Substituting, we get

$$f(1) = y = 2(1) - 3 = -1$$

Therefore, $(1, f(1)) = (1, -1)$.   ❖

Linear depreciation is an example of modeling where the value of an item is given as a linear function of the time the item is in use.

Example 1.8   Suppose you are the owner of a small business in need of a copying machine. The machine you have decided to buy for the company costs \$45,000 now, and has an expected scrap value of \$3,000 in 6 years. Find the linear function that models the value of the machine as a function of time.

Solution
Time starts when we purchase the machine. This gives us the ordered pair (0, 45,000). In six years, the machine will be worth \$3,000. This gives us the ordered pair (6, 3,000). Since we have two points, we can find the slope. The magnitude of the slope is the **rate of depreciation**.

$$m = \frac{45,000 - 3,000}{0 - 6} = \frac{42,000}{-6} = -7,000$$

The slope of the linear model is $-7,000$ and thus, the rate of depreciation of the machine is $7,000 per year. We can see that we are given the y-intercept $(0, 45,000)$. Therefore, using the slope-intercept form of a line we find that the value of the machine as a function of time is given by $f(x) = -7,000x + 45,000$. ❖

Supply and demand are examples of mathematical modeling for the price of an item as a function of the number of items available in the market. The **supply curve** represents an equation modeling the willingness of a manufacturer to supply certain quantities of a product at different prices. The **demand curve** represents an equation modeling the willingness of consumers to buy certain quantities of a product at different prices. The **intersection** point (where the lines cross) of the linear models for supply and demand is called the **market equilibrium**. This is an important concept that you will learn more about in economics.

Example 1.9   We are told that the demand equation for certain computer chips is given by $D(x) = -6x + 13$, where x is in thousands. The supply equation for these computer chips is $S(x) = 4x + 3$, where x is also in thousands of computer chips. What is the market equilibrium?

Solution
The supply and demand equations are graphed in Figure 1.8. Notice that the functions are graphed only in the first quadrant — both the number of items and the price must be non-negative!

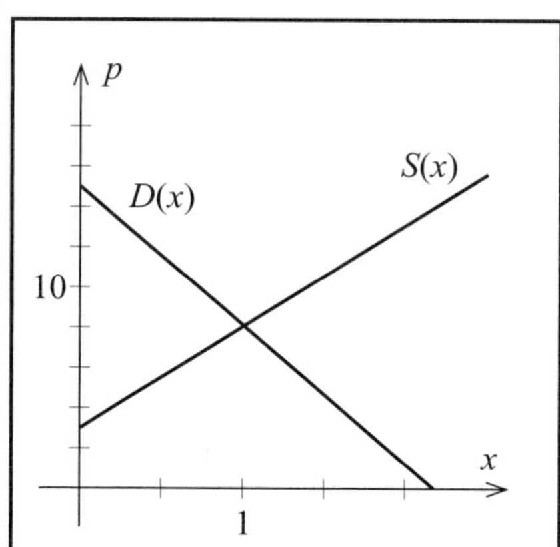

Figure 1.8   Graph of Supply and Demand

At the market equilibrium, both equations have the values (x_0, p_0) and so

$$p_0 = -6x_0 + 13$$

and

$$p_0 = 4x_0 + 3$$

Since these are the same price, p_0, we can set them equal to each other to find x_0.

$$-6x_0 + 13 = 4x_0 + 3$$
$$-10x_0 = -10$$
$$x_0 = 1$$

We can use this value in either the supply or demand equation to find p_0

$$D(x_0) = p_0 = -6(1) + 13 = 7$$
$$S(x_0) = p_0 = 4(1) + 3 = 7$$

Alternatively, we can graph these two equations together on a TI-83 graphing calculator and use the built-in utility **[2ⁿᵈ][CALC]** $5:intersect$ to find the intersection is at $(1, 7)$.

Since the intersection point of these two equations is (1, 7), that is the market equilibrium. That means the market equilibrium quantity is 1,000 computer chips (since x was given in thousands) and the market equilibrium price is \$7 per computer chip.   ❖

 **Note:** *You should now complete Activity E in the Lines and Slopes Module.*

 ## FURTHER EXPLORATIONS

For further explorations in the area of lines and slopes, please visit

http://www.finitemathtutor.com/explore/chapter1/

 ## EXERCISES

Exercise 1.1   Graph a horizontal line that passes through the point (1, –1).

Exercise 1.2   Graph a vertical line that passes through the point (–2, –3).

Exercise 1.3   Graph the line $x = -5$ and a line parallel to it that passes through the point (2, 4).

Exercise 1.4   Graph the line $y = 4$ and a line perpendicular to it that passes through the point (0, –2).

Exercise 1.5   Find a value for a such that the line passing through the points (12, 1) and (10, a) is parallel to the line passing through the points (0, 10) and (6, $a - 7$).

Exercise 1.6   Find the equation of the line that passes through the point (2, –1) and is perpendicular to the line $y = cx - 1$.

Exercise 1.7   Where do the lines $x = 3$ and $y = -2x + 6$ intersect?

Exercises 1.8 and 1.9 use the following information:

A certain candy manufacturer produces boxes of a candy called Choco Droplets. They are willing to supply 200 boxes at a market price of $1 per box. They are not willing to supply any boxes of Choco Droplets at a market price of $0.75 or less. Research indicates that consumers will purchase 150 boxes of Choco Droplets if they are $1.25 per box, but only 25 boxes at a market price of $1.75 per box.

Exercise 1.8   What is the demand equation for Choco Droplets?

Exercise 1.9   What is the supply equation for Choco Droplets?

Exercise 1.10   The demand for Super-Duper TVs is given by $-13x + 27{,}325$ and the supply for Super-Duper TVs is given by $6x + 1{,}257$. What is the equilibrium quantity and price of Super-Duper TVs? x is the number of TVs and the units for supply and demand are dollars.

SAMPLE QUIZ

Question 1.1   Graph a horizontal line that passes through the point $(-1, -2)$.

Question 1.2   Graph a vertical line that passes through the point $(3, -1)$.

Question 1.3   Graph the line $x = 2$ and a line parallel to it that passes through the point $(-1, 2)$.

Question 1.4   Graph the line $y = -1$ and a line perpendicular to it that passes through the point $(-2, 3)$.

Question 1.5   Find a value for a such that the line passing through the points $(2, 1)$ and $(6, a)$ is parallel to the line passing through the points $(-1, 4)$ and $(5, a + 2)$.

Question 1.6   Find the equation of the line that passes through the point $(0, 1)$ and is perpendicular to the line $y = kx + 4$.

Question 1.7   Where do the lines $x = -2$ and $y = 5x - 3$ intersect?

Question 1.8   A coffee shop sells gourmet biscotti (Italian cookies). When the price of biscotti is $4 each, they sell two of them. When the price is $2, they sell 20 biscotti. Find the demand equation for these biscotti at this shop.

Question 1.9   A supplier of decorative scarecrows is willing to make 12 scarecrows when she can sell them at $40 each. For every $10 increase in the selling price, she will make 6 more scarecrows to sell. Find the supply equation for these scarecrows.

Question 1.10   The demand for fuel in Utopia is given by $3x + 6p - 30 = 0$ and the supply for fuel is given by $4x - 5p + 12 = 0$. The price p is given in Utopian bucks and the demand x is given in millions of liters. What is the equilibrium price? (Round to two decimal places.) How many liters will be sold at the equilibrium price?

Least Squares

2

In Chapter 1, we learned that two points uniquely determine a straight line. A straight line has a constant slope and intercept. If there are more than two data points, it is not possible in general to find a single line passing through all the points. The **least-squares method** provides a way of fitting a straight line to more than two data points. It does so by making the line as close as possible to all the points. The slope and the intercept of this "best fitting" line are given as a simple formula in terms of the x- and y-values of the data. The process of fitting a line to data is called **linear regression**.

2.1   LINEAR REGRESSION

It is customary to plot data in two variables as a set of points (ordered pairs) with labeled axes, tick marks, and a suitable scale. The resulting graph is called a **scatter plot**. Suppose you are given the set of points $\{(0, 0), (1, 2), (2, 4), (3, 6)\}$ to plot. It is fairly obvious that these points come from the function $y = 2x$. Each of the points is exactly on the line $y = 2x$ as shown in Figure 2.1. Suppose, however, the data is just slightly different: $\{(0, 0), (1, 2), (2, 5), (3, 6)\}$. Only the third point is changed, shown in Figure 2.2.

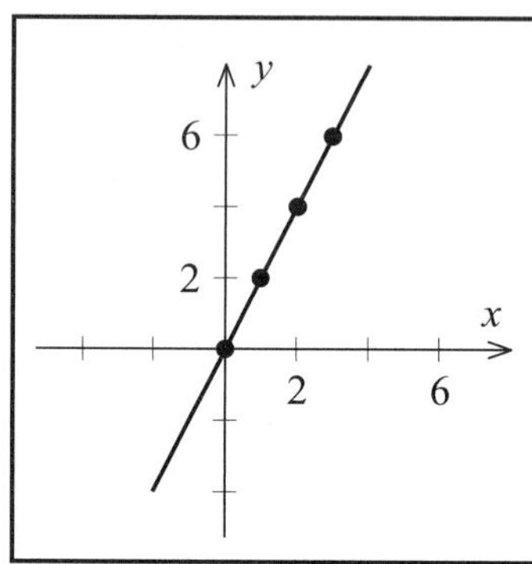

Figure 2.1   Linear Data

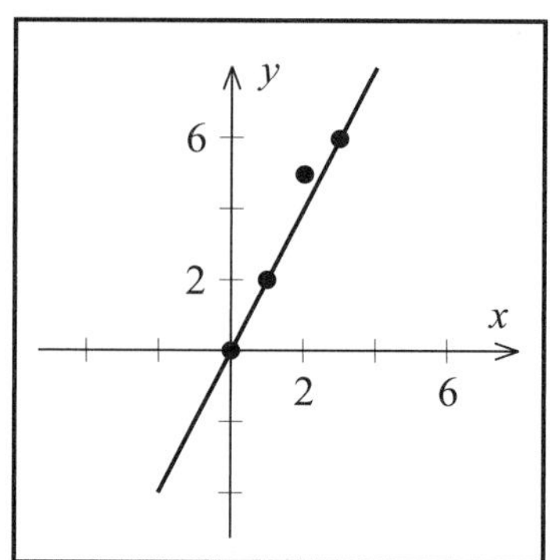

Figure 2.2   Non-Linear Data

Are we willing to say that the data still comes from the line $y = 2x$? Since data in the world often comes with error, this is a very real question. Only one point seems to be off the straight line, so we might be willing to discount it as a result of measurement error, an **outlier**. This involves a statistical description of the data, which is discussed in Chapter 8.

Suppose we don't know where the errors occur, or even the magnitude of the possible errors. All we have are the points themselves, and the assumption that the points come from a linear function. We are led to the fundamental question:

What is the *best possible* linear function that fits the data we are given?

Clearly, we have no problem if there are only two points, since two points determine only one possible straight line. The problems come with more than two data points. Suppose we are given three points. *Three* possible lines go through the various sets of points (Figure 2.3). If there are more points, there are even more possible lines. Which of these lines is best? What do we mean by best?

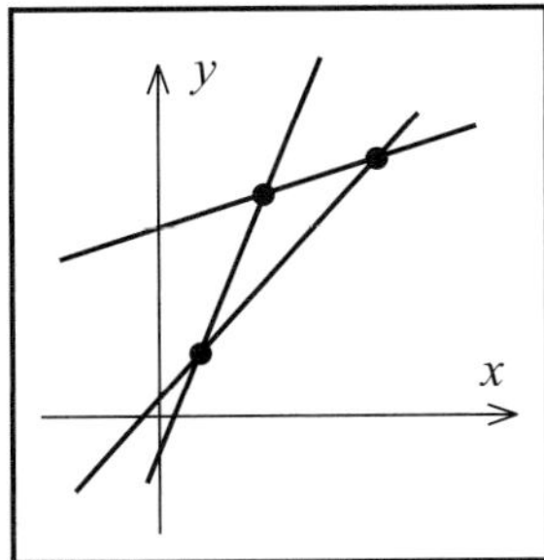

Figure 2.3   Three Possible Lines Connecting Pairs of Points

One possible way out of this dilemma is to quantify what we mean by "best." An intuitive understanding of "best" comes from the notion of error. By **error**, we don't mean the unknown error that is present in the data, but the error that comes from attempting to describe the data as coming from a single straight line.

 **Note:** *You should now complete Activity A in the Least-Squares Module.*

The equation of the single line describing the data is of the form $y = mx + b$, where m and b are as yet unknown. The first data point is (x_1, y_1), and the value of the line at this point is $y_1 = mx_1 + b$ (Figure 2.4). The difference between the actual data point and the value that the line predicts is $|(mx_1 + b) - y_1|$.

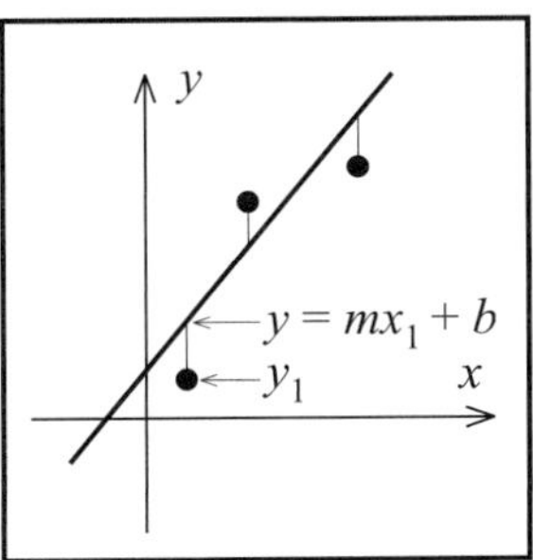

Figure 2.4   A Line with Errors

Each data point contributes to this "error" and the sum of these errors is

$$|(mx_1 + b) - y_1| + |(mx_2 + b) - y_2| + \cdots$$

We would like to make this sum as small as possible. This turns out to be mathematically challenging, so we modify our notion of error slightly. Instead, we will make the sum of the squares (SOS) of the errors as small as possible.

$$\text{SOS} = |(mx_1 + b) - y_1|^2 + |(mx_2 + b) - y_2|^2 + \cdots$$

Note: *You should now complete Activity B and Least-Squares Exercise 1 in the Least-Squares Module.*

Rather than rely on our ability to minimize this error by moving the line, we want to find a mathematical procedure (algorithm) which will do this for us, exactly the same way, and for any possible sets of data. This procedure is called the method of least squares.

2.2   METHOD OF LEAST SQUARES

Given N data points $(x_1, y_1), (x_2, y_2), \ldots, (x_N, y_N)$, the optimal values for m and b are given by the solution of the two by two system of equations:

$$m\left(\sum_{i=1}^{N} x_i^2\right) + b\left(\sum_{i=1}^{N} x_i\right) = \sum_{i=1}^{N} x_i y_i$$

$$m\left(\sum_{i=1}^{N} x_i\right) + bN = \left(\sum_{i=1}^{N} y_i\right)$$

It is very useful to note that a two by two system, in terms of the variables m and b, of the form

$$A_{11}m + A_{12}b = C_1$$
$$A_{21}m + A_{22}b = C_2$$

has the solution

$$m = \frac{A_{21}C_1 - A_{11}C_2}{A_{21}A_{12} - A_{11}A_{22}}$$

$$b = \frac{A_{22}C_1 - A_{12}C_2}{A_{11}A_{22} - A_{21}A_{12}}$$

if $A_{11}A_{22} - A_{11}A_{22} \neq 0$. The line $y = mx + b$, using the optimal values of m and b above, is called the **regression line**.

In the Least-Squares Module, the Least-Squares Applet allows you to insert points anywhere in a drawing area and automatically recalculates the slope each time you add a point or move a point. In Figure 2.5 we can see how adding additional points changes the slope (and intercept) of the least-squares line. The lines that connect the data points to the least-squares line show the error.

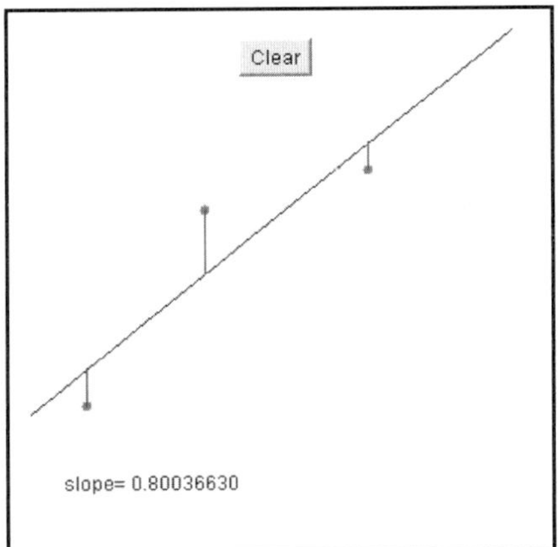

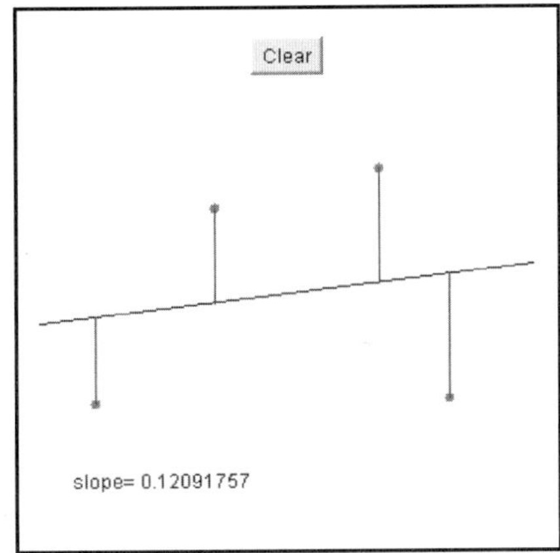

Figure 2.5   Examples of the Interactive Least-Squares Applet

Figure 2.6 shows five data points and two lines. The line with the error bars drawn is a guess for the least-squares line. The other line is the calculated least-squares line. In the Least-Squares Module you are given the opportunity to test how well you can guess the position of the regression line for five data points.

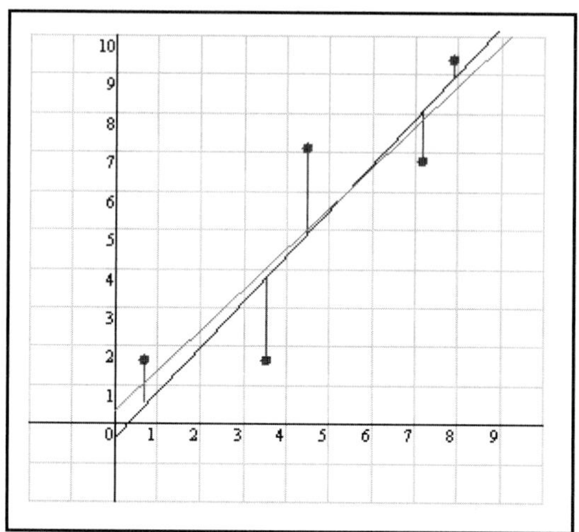

Figure 2.6 Least-Squares Applet Regression Line

 **Note:** *You should now complete the Interactive Tutorial in the Least-Squares Module.*

 **Example 2.1**   Given the following national sales information for ink-jet printers, find the linear demand equation that best fits the data (the regression equation). The sales figures are in millions of dollars and the price of the printer is in dollars.

Sales (x)	2.3	2.1	1.8	1.5
Price (y)	149.99	159.99	175.00	199.99

Solution

In this case the two by two system of equations reduces to

$$15.19m + 7.7b = 1295.941$$
$$7.7m + 4b = 684.97$$

which has the solution

$$m = -61.5680$$
$$b = 289.7609$$

The demand equation will then be the regression line

$$y = D(x) = -61.5680x + 289.7609$$

In many cases it is easier to use a graphing calculator (such as the TI-83) that has built-in support for linear regression, for example using the **[STAT][CALC]** `4:LinReg(ax+b)` sequence. The Least-Squares Applet will also accept new data and find the regression equation and its graph, as shown in Figure 2.7.

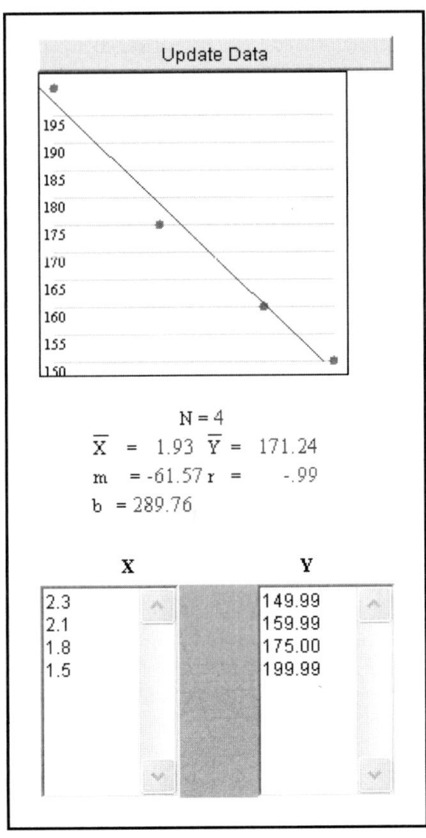

Figure 2.7   Least-Squares Applet

One measure for how closely data can be fit to a straight line is r, the **correlation coefficient**. The value of r is a number between -1 and $+1$ and is a statistical measure of the linear relationship between two sets of data. The closer the absolute value of the correlation coefficient is to $+1$ in *magnitude*, the more the data lies on a straight line. If you find $|r| = 1.0$, then the data can be fit exactly with a straight line. However, a coefficient of $r = 0$ effectively means there is no linear relationship between the data. When a graphing calculator or spreadsheet is used to find the regression equation, the correlation coefficient, r, is given along with the values of m and b.

 **Example 2.2** A pre-calculus instructor gives his class a pretest to determine their mathematical skills coming into his class. At the end of the semester, he records the final exam scores. He then compares the two scores and looks for a relationship between them. The pretest scores are the x-data and the final exam scores are the y-data for the 19 students. Find the regression equation. Graph the data and the regression line. Do you think the regression line is a good predictor of performance on the final exam?

Pretest (x)	57	65	34	78	65	86	45	92	55	73	27	85	80	33	68	78	43	66	76
Final (y)	68	76	58	79	70	80	60	95	50	68	49	76	95	73	80	72	55	80	45

Solution

The regression line is given by $y = mx + b = 0.45712x + 40.93222$. In Figure 2.8 you can see that while the regression line roughly gives the trend of the data (final exam scores are higher if the pretest scores are higher), a number of data points lie well below (and above) the line. The correlation coefficient is 0.62137, which would indicate that the pretest grade is only a moderately good predictor of the final exam score.

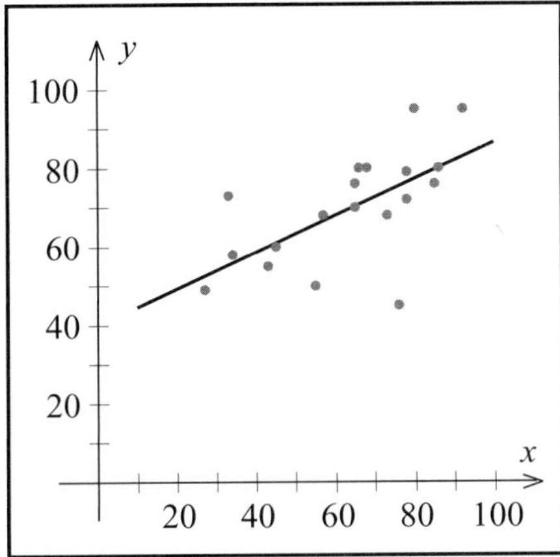

Figure 2.8 Exam Data and Regression Line ❖

Example 2.3 Suppose data comes from points on the curve $y = x^2$, in the range from $x = -3$ to $x = 5$. What is the formula for the regression line, and what is the correlation coefficient?

x-values	−3	−2	−1	0	1	2	3	4	5
y-values	9	4	1	0	1	4	9	16	25

Solution

The regression line is given by $y = 2x + 5.666$. The correlation coefficient is only 0.66178. This is an important point. If the data does not come from a linear function, then linear regression will not necessarily provide good results! ❖

2.3   NON-LINEAR REGRESSION

Clearly, if the data is known to come from a quadratic function, then finding a quadratic curve that minimizes the sums of squares error is a better approach. This is called **quadratic regression**, and can be done on the TI-83 using the **[STAT][CALC]** 5:QuadReg sequence. A quadratic regression calculation on the values from Example 2.3 yields the curve

$$y = ax^2 + bx + c$$

with $a = 1$, $b = 0$, $c = 0$, and a correlation coefficient of 1.

Given a set of data, one does not always know what kind of function to fit. Should you use a linear function $y = mx + b$, a quadratic function $y = ax^2 + bx + c$, or a cubic, an exponential, or some other function?   In the linear case, only two constants (m and b) are undetermined. The linear regression problem is said to have two **degrees of freedom**. In the quadratic case, there are three constants (a, b, and c), so it has three degrees of freedom. The cubic has four degrees of freedom, and so on. The higher the degree of the polynomial, the more the degrees of freedom, and presumably the better the fit. This is often the case, but a higher degree can lead to problems. See [Chapter 2, Explorations] for a case where approximating the data with a high-degree polynomial leads to bad behavior. This topic is taken up again in section 10.5 (Trend Analysis).

The TI-83 calculator has several regression methods available:

- LinReg(ax+b) – fitting a linear function $y = ax + b$ to the data
- QuadReg – fitting a quadratic function $y = ax^2 + bx + c$ to the data
- CubicReg – fitting a cubic function $y = ax^3 + bx^2 + cx + d$ to the data
- QuartReg – fitting a quartic function $y = ax^4 + bx^3 + cx^2 + dx + e$ to the data
- LnReg – fitting a logarithmic function $y = a + b \ln x$ to the data
- ExpReg – fitting an exponential function $y = ab^x$ to the data
- PwrReg – fitting a power function $y = ax^b$ to the data

A few other regression methods are available on the TI-83, but are beyond the scope of this module. In any case, one can try various functions to the data, and find out which has the highest correlation coefficient. The function with the highest correlation has the best fit, but is not always the best model. When using the regression function for data outside the given x range, you must be careful that the extrapolated behavior is reasonable.

What if the data comes from a more complex set of functions, such as a combination of powers, exponential functions, and trigonometric functions? In this case, we try to fit a function of the form

$$y = c_1 f_1(x) + c_2 f_2(x) + c_3 f_3(x) + \cdots$$

where the functions are known, and try to determine the coefficients $c_1, c_2, c_3, \ldots$. In this case, it is often a matter of intuition and good luck to find a set of functions with few degrees of freedom that fits the data well.

Turn to the Non-Linear Least-Squares Applet [Chapter 2, Explorations] and enter in the quadratic data from Example 2.3. Try various functions (such as x^3 or ab^x) to see how good the "fit" is.

2.4   DERIVATION OF LEAST SQUARES [OPTIONAL]

Our goal is to minimize the sum of squares error function

$$S = S(m,b) = \sum_{i=1}^{N} \left| y_i - (mx_i + b) \right|^2$$

by a special choice of the constants m and b.

First, we show that this is a *quadratic expression* in terms of m and b. To do this, expand this expression in terms of m and b, which gives

$$S(m,b) = \sum_{i=1}^{N} \left[y_i^2 - 2y_i(mx_i + b) + (mx_i + b)^2 \right]$$

$$= \sum_{i=1}^{N} \left[y_i^2 - 2y_i(mx_i + b) + m^2 x_i^2 + 2mbx_i + b^2 \right]$$

$$= m^2 \left(\sum_{i=1}^{N} x_i^2 \right) + m \left(-2\sum_{i=1}^{N} x_i y_i \right) + b^2 \left(\sum_{i=1}^{N} 1 \right) + b \left(-2\sum_{i=1}^{N} y_i \right) + mb \left(2\sum_{i=1}^{N} x_i \right) + \sum_{i=1}^{N} y_i^2$$

$$= Am^2 + Bm + Cb^2 + Db + Emb + F$$

which is indeed a quadratic expression in terms of m and b. The constants $A, B, C, D, E,$ and F are given by

$$A = \sum_{i=1}^{N} x_i^2, \qquad B = -2\sum_{i=1}^{N} x_i y_i, \qquad C = \sum_{i=1}^{N} 1 = N$$

$$D = -2\sum_{i=1}^{N} y_i, \qquad E = 2\sum_{i=1}^{N} x_i, \qquad F = \sum_{i=1}^{N} y_i^2$$

Now, suppose m and b actually minimize this sum. Then for any choice of constants g and h, we must have

$$S(m + g, b + h) \geq S(m, b)$$

since $S(m, b)$ is the minimum. If it were otherwise, we would use the new expression as the minimum! Expanding the left-hand side of this inequality and then subtracting the right-hand side from the left-hand side, we get

$$A(2mg + g^2) + Bg + C(2bh + h^2) + Dh + E(gb + mh + gh) \geq 0$$

or

$$g(2mA + B + Eb) + h(2Cb + D + Em) + Ag^2 + Ch^2 + Egh \geq 0$$

If this inequality is to be true for any *arbitrary* g and h, then the terms in the parentheses must vanish. Why is this? Since it is true for *any* g and h, then let $g = 0$. We must have the inequality

$$h(2Cb + D + Em) + Ch^2 \geq 0$$

for *any* h. Suppose $(2Cb + D + Em)$ is negative. Take h to be positive, and divide by h. We then have (for all positive h)

$$(2Cb + D + Em) + Ch \geq 0$$

Even if C is positive, if we make h small enough, Ch won't be enough to cancel out the first term, and the inequality is violated.

Similarly, suppose $(Cb + D + Em)$ is positive. Make h negative, and divide by h (reversing the sign of the inequality):

$$(2Cb + D + Em) + Ch \leq 0$$

If h is small enough, then Ch cannot compensate for the first term, and the inequality is violated. The only way out is to have the quantity $(2Cb + D + Em)$ vanish! If we set h equal to zero, by a similar argument, $(2mA + B + Eb)$ must vanish! This kind of argument is the basis for a branch of mathematics called the **calculus of variations**.

FURTHER EXPLORATIONS

For further explorations in the area of Least Squares please visit the URL

http://www.finitemathtutor.com/explore/chapter2/

EXERCISES

Exercises 2.1 and 2.2 use the following data set:

x	0	1	2	3	3	4	5
y	3	4	6	8	9	9	12

Exercise 2.1   Find the equation of the least-squares line for the given data. (Round each term to four decimal places.)

Exercise 2.2   Draw a scatter plot for the given data and graph the least-squares line.

Exercises 2.3–2.5 use data from the Delightful Donut shop below.

The Delightful Donut shop has tracked the sales of a new item, cinnamon rolls. The number of rolls sold each month is shown in the following table:

Month (x)	Jan.	Feb.	Mar.	Apr.	May	June
Number Sold (y)	20	32	41	53	60	66

Exercise 2.3   Find an equation of the least-squares line for this data. (Let $x = 1$ represent January.)

Exercise 2.4   Use the least-squares line to estimate the sales in August.

Exercise 2.5   Do you think that the linear model will be useful in predicting sales the next year?

Exercises 2.6–2.9 use the data shown below:

Six people were asked their annual income (in thousands of dollars) and the average number of hours they spend watching television each week. The data collected is shown in the following table.

Income (x)	10	27	38	55.5	70	100
Hours Watching TV (y)	50	33	28	20	12	6

Exercise 2.6　Find the equation of the least-squares line for this data.

Exercise 2.7　Using the least-squares equation, estimate the average number of hours of television watched per week by a person with an annual income of $45,800.

Exercise 2.8　Using the least-squares equation, estimate the annual income (to the nearest dollar) of a person who watches an average of 10 hours of television per week.

Exercise 2.9　The correlation coefficient for this data is –0.9625. Does this suggest that the data is linear in nature?

Exercise 2.10　If the scatter plot of a set of data points reveals that the points form a square, will the correlation coefficient be closer to –1, 0, or 1?

SAMPLE QUIZ

Questions 2.1–2.4 use the information below:

Five people are surveyed about their credit card debt. The table below shows the amount owed on January 1 and the person's annual income (both in dollars):

Annual Income (x)	23,000	31,000	38,000	48,000	51,000
Credit Card Balance (y)	1,200	1,500	1,300	2,000	2,400

The least-squares equation is $y = 0.039\, x + 200.70$.

Question 2.1　Why do we use the least-squares method for this data?

Question 2.2　Use the least-squares equation to estimate the credit card debt for a person with an annual income of $40,000.

Question 2.3　Use the least-squares equation to estimate the annual income of a person with a credit card debt of $1,900.

Question 2.4　The correlation coefficient for this data is $r = .890$. What does this tell you about our model?

Questions 2.5–2.8:

You wish to test a new gram scale for accuracy. You know that a penny weighs 3 grams, so you test the new scale by weighing pennies. Your results are

Number of Pennies (x)	5	10	20
Weight in Grams (y)	14	31	63

Question 2.5   What is the equation of the least-squares line?

Question 2.6   What would the equation of the least-squares line be if the scale were perfectly accurate?

Question 2.7   Use the least-squares equation to estimate the weight of 12 pennies.

Question 2.8   Graph the data and the least-squares line. What can you say about the accuracy of the scale?

Question 2.9   When the correlation coefficient for a linear regression calculation for a data set is very close to 1 or –1, what can you say about using a linear model for this data?

Question 2.10   When the correlation coefficient for a linear regression calculation for a data set is very close to 0, what can you say about using a linear model for this data?

Matrices and Linear Systems

In Chapter 1, we learned that in two dimensions, two non-parallel lines intersect at a single point. We can consider the two lines to be a **system of linear equations**. If each of these lines is described by a linear equation, then the point of intersection is the solution of the system of two linear equations. What happens when we have more equations, and more independent variables?

3.1 SYSTEMS OF LINEAR EQUATIONS

A system of linear equations is two or more linear equations to be solved simultaneously. We begin by looking at some examples of linear systems that have two equations and two unknown variables.

In general, there are three possible solution situations for two linear equations with two variables:

- One solution (one intersection point, Figure 3.1)
- No solution (no intersection, parallel lines, Figure 3.2)
- Infinitely many solutions (infinitely many points of intersection, the two lines coincide, Figure 3.3)

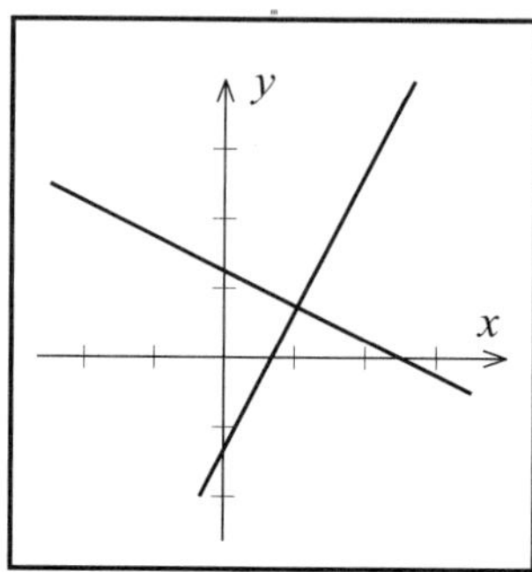

Figure 3.1   Two Linear Equations – One Solution

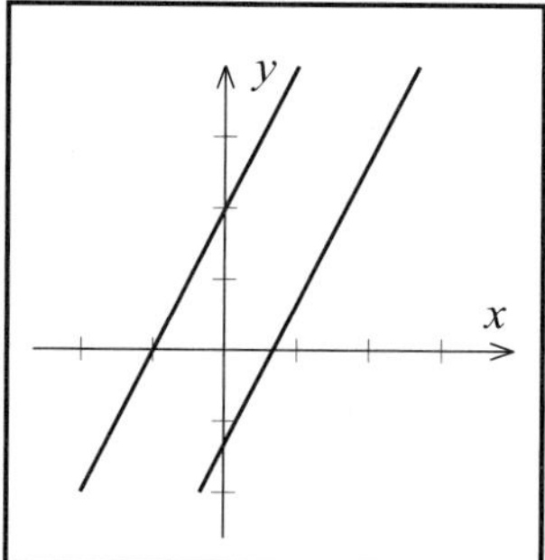

Figure 3.2   Two Linear Equations – No Solution

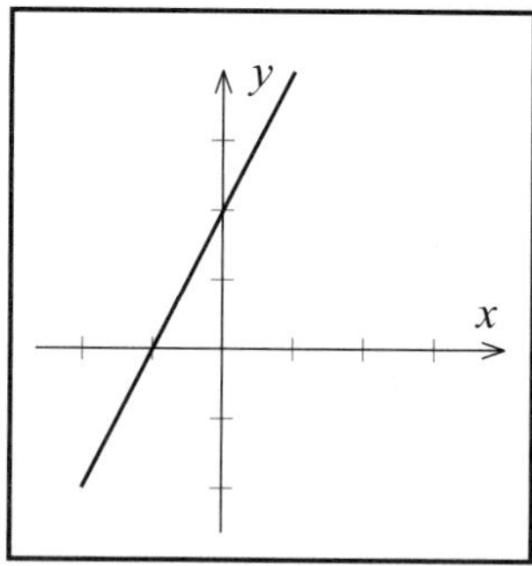

Figure 3.3   Two Linear Equations – Infinitely Many Solutions

Example 3.1 Find the solution to the system

$$2x - y = 1$$
$$3x + 2y = 12$$

Solution

Begin by graphing the lines as shown in Figure 3.4. The solution is the place where both equations are true at the same time. We can label this point as (x_0, y_0). We can find the values of x_0 and y_0 algebraically or with a graphing calculator.

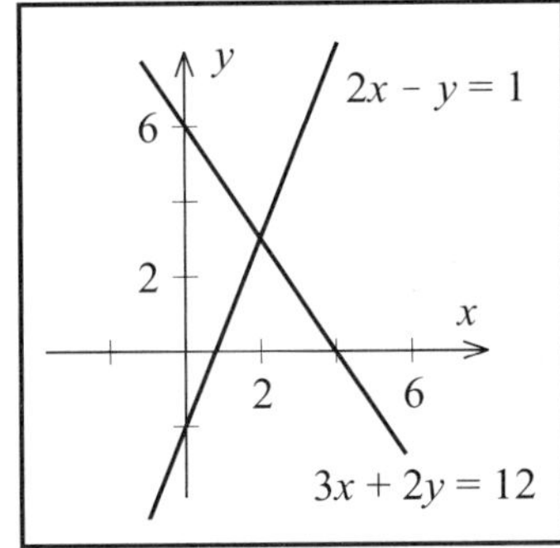

Figure 3.4 Graph of Example 3.1

Algebraic method: Start by writing each equation in slope-intercept form. For Equation 1, $2x - y = 1$, we can solve for y by first isolating y,

$$-y = -2x + 1$$

and then multiplying both sides by -1

$$y = 2x - 1$$

For Equation 2, $3x - 2y = 12$, we can solve for y by first isolating y,

$$2y = -3x + 12$$

and then dividing both sides by 2

$$y = -\frac{3}{2}x + 6 = -1.5x + 6$$

At the intersection point both lines must have the same y-value, y_0, so we are able to solve for the x-value of the intersection,

$$2x_0 - 1 = -1.5x_0 + 6$$
$$3.5x_0 = 7$$
$$x_0 = 2$$

The y-value of the solution can be found from either of the slope-intercept lines.

$$y_0 = 2(2) - 1 = 3$$
$$y_0 = -1.5(2) + 6 = 3$$

This must give us the same value, or an *error has occurred*. The solution for the system is the point (2, 3).

Calculator method: The intersection of two lines can be found using a graphing calculator. Using the TI-83, once the two equations are entered into the [Y=] screen, a suitable window must be found. The window must show the intersection point; for this example the using the window Xmin = 0, Xmax = 5, Ymin = 0, Ymax = 5 will work. Use the [2ⁿᵈ][CALC] $5:intersect$ utility to find the values at the intersection point.   ❖

Alternatively, we can use the **method of substitution**, solving for one variable at a time and substituting that value into the remaining equations, to find the solution to a linear system.

Example 3.2   Find the solution to the system

$$4x + 3y = 11$$
$$2x - 5y = -1$$

Solution
Solve the first equation for y, since both equations have y-terms.

$$4x + 3y = 11$$

$$3y = -4x + 11$$

$$y = -\frac{4}{3}x + \frac{11}{3}$$

Now, if we substitute this identity into the second equation of the system, we obtain

$$2x - 5\left(-\frac{4}{3}x + \frac{11}{3}\right) = -1$$

We can simplify this expression and then solve for x, through the following procedure:

$$2x + \frac{20}{3}x - \frac{55}{3} = -1$$

Find the common denominator, and isolate the variable x.

$$\frac{6}{3}x + \frac{20}{3}x = \frac{55}{3} - \frac{3}{3}$$

Collecting terms,

$$\frac{26}{3}x = \frac{52}{3}$$

then solving for x, gives us

$$x = \frac{52}{3} \cdot \frac{3}{26} = 2$$

Now, substitute $x = 2$ into the first equation where we solved for y.

$$y = -\frac{4}{3}(2) + \frac{11}{3} = -\frac{8}{3} + \frac{11}{3} = \frac{3}{3} = 1$$

Our solution is (2, 1) and represents one unique point in the plane (Figure 3.5).

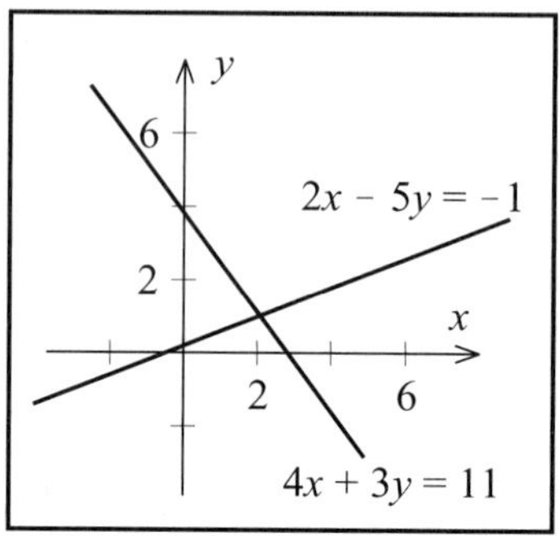

Figure 3.5   Graph of Example 3.2   ❖

 **Example 3.3**   Find the solution to the system

$$3x - y = 10$$
$$6x - 2y = 5$$

Solution

Once again, solve the first equation for y.

$$3x - y = 10$$
$$-y = -3x + 10$$
$$y = 3x - 10$$

Now, substitute for y in the second equation.

$$6x - 2(3x - 10) = 5$$
$$6x - 6x + 20 = 5$$
$$20 = 5$$

What went wrong? Since $20 \neq 5$, this system has *no solution* and graphs as two parallel lines (Figure 3.6).

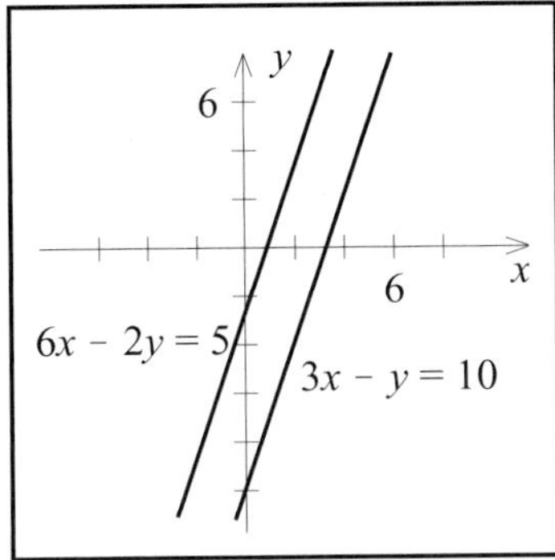

Figure 3.6   Graph of Example 3.3

❖

Example 3.4   Find the solution to the system

$$4x + 6y = 4$$
$$2x + 3y = 2$$

Solution

Begin by solving for y in the second equation

$$2x + 3y = 2$$
$$3y = -2x + 2$$
$$y = -\frac{2}{3}x + \frac{2}{3}$$

Substituting this identity into the first equation gives

$$4x + 6\left(-\frac{2}{3}x + \frac{2}{3}\right) = 4$$
$$4x - 4x + 4 = 4$$
$$4 = 4$$

The statement $4 = 4$ is always true. When these two lines are graphed in Figure 3.7 we find only one line.

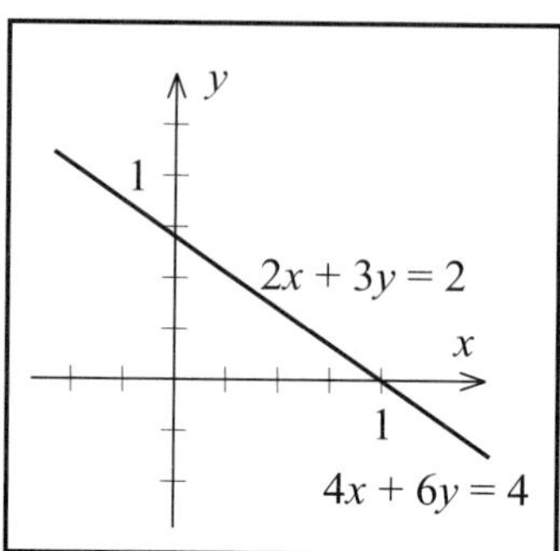

Figure 3.7   Graph of Example 3.4

This system has infinitely many solutions. We can write down what those solutions look like by finding the coordinates of the points on the line as follows:

$$(x, y) = \left(x, -\frac{2}{3}x + \frac{2}{3} \right)$$

where x is *any* **real number**. This tells us that all solutions look like the above ordered pair whenever we choose any real number for x. If our favorite value of x is 0, then the solution for that choice of x is $(0, \frac{2}{3})$. When a value for x is chosen and the corresponding y value is found, the solution is called a particular solution. If we have infinitely many choices for x to substitute into the ordered pair, we will have infinitely many solutions. ❖

 **Note:** *You should now complete Activity A in the Matrices Module.*

A linear system of three equations with three unknowns has a similar representation in terms of the number of general solution situations (Figures 3.8–3.10), but clearly cannot be graphed in a plane to find the intersection (or lack of intersection) as a possible solution method.

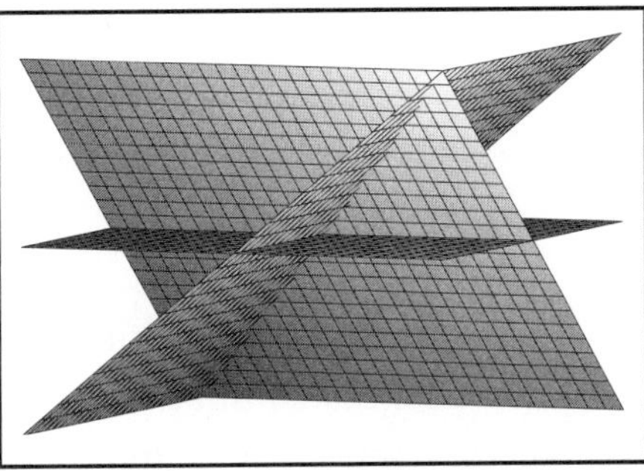

Figure 3.8 Exactly One Solution in Three Dimensions

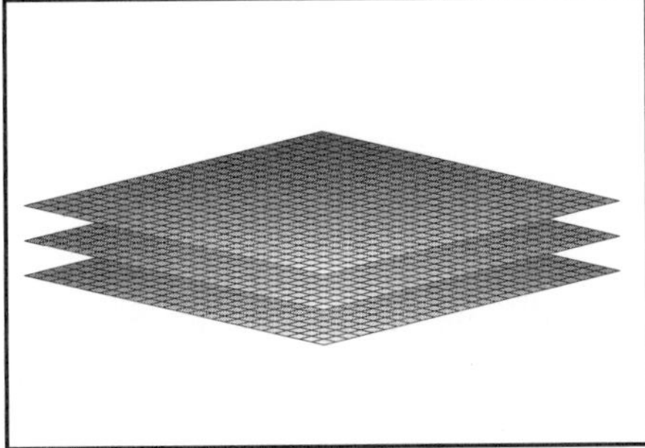

Figure 3.9 No Solution in Three Dimensions

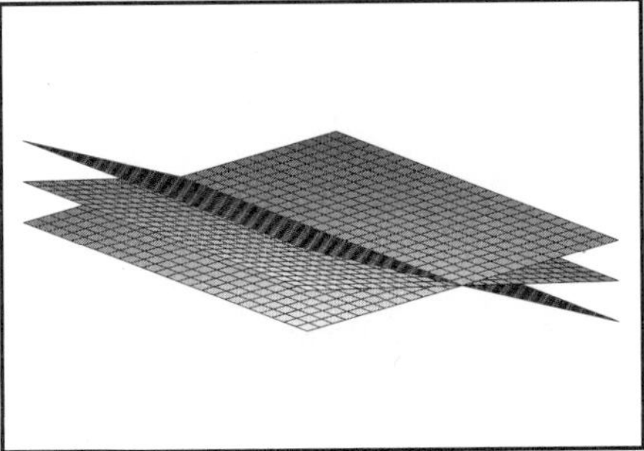

Figure 3.10 Infinitely Many Solutions in Three Dimensions

An example of a system with three equations and three variables is

$$2x + 4y + 5z = 4$$
$$x - 2y - 3z = 5$$
$$x + 3y + 4z = 1$$

Graphing is not a practical way to solve these systems since most graphing calculators do not have three-dimensional capabilities. Substitution is quite complicated and becomes even more so when there are more variables. An example of a system of four equations with four variables is

$$2x + 4y + 5z - w = 4$$
$$x - 2y - 3z + 4w = 5$$
$$x + 3y + 4z - 2w = 1$$
$$-x - y - 2z + 3w = 2$$

Since real life doesn't restrict us to the 26 letters of the alphabet for the number of possible variables, changing the variable notation to the use of subscripts (that is, x_1, x_2, x_3, x_4) enables us to write an equation in n variables. Since there are infinitely many positive integers to use for subscripts, we can accommodate any size system with this notation. Adopting the new notation, the above system becomes

$$2x_1 + 4x_2 + 5x_3 - x_4 = 4$$
$$x_1 - 2x_2 - 3x_3 + 4x_4 = 5$$
$$x_1 + 3x_2 + 4x_3 - 2x_4 = 1$$
$$-x_1 - x_2 - 2x_3 + 3x_4 = 2$$

As you can see, using substitution to solve such large systems would be an overwhelming and unpleasant task that involves writing these equations over and over, creating a lot of extra baggage, the variables. To make finding the solution easier, we organize the data (the coefficients of the variables) into blocks of real numbers called **matrices**.

.2 MATRICES

Matrices are blocks or arrays of real numbers. Matrices have dimensions given as the number of rows by the number of columns. So, referring to an $M{\times}N$ matrix means that the matrix has M rows and N columns. Some examples of matrices of different types and dimension are given below.

A **square matrix** has the number of rows equal to the number of columns. A $3{\times}3$ square matrix looks like

$$\begin{bmatrix} 2 & 4 & 5 \\ 1 & -2 & -3 \\ 1 & 3 & 4 \end{bmatrix}$$

Matrix **A** is a $3{\times}1$ **column matrix** since it has three rows and one column. Matrix **B** is a $1{\times}3$ **row matrix** since it has one row and three columns. Matrix **C** is called a **zero matrix** since every entry is the number 0. It is also a square $2{\times}2$ matrix. Matrix **I** is the $2{\times}2$ square **identity matrix**. The identity matrix will always be denoted as I.

$$A = \begin{bmatrix} 2 \\ 1 \\ 1 \end{bmatrix} \qquad B = \begin{bmatrix} 2 & 1 & 1 \end{bmatrix} \qquad C = \begin{bmatrix} 0 & 0 \\ 0 & 0 \end{bmatrix} \qquad I = \begin{bmatrix} 1 & 0 \\ 0 & 1 \end{bmatrix}$$

Two matrices are equal (**matrix equality**) if and only if they have the same dimensions and the corresponding entries are equal. Two examples of what matrix equality is *not* are shown below:

$$\begin{bmatrix} 1 & 3 & 5 \\ 2 & 4 & 6 \end{bmatrix} \neq \begin{bmatrix} 1 & 2 \\ 3 & 4 \\ 5 & 6 \end{bmatrix} \qquad \begin{bmatrix} 2 & 3 \\ 4 & 6 \end{bmatrix} \neq \begin{bmatrix} 2 & 3 \\ 4 & 7 \end{bmatrix}$$

Two matrices can be added (**matrix addition**) if and only if they have the same dimensions. If they do have the same dimensions, elements in the same positions are added and are placed in the corresponding position in the final matrix, as shown:

$$\begin{bmatrix} 2 & 3 \\ 4 & 6 \end{bmatrix} + \begin{bmatrix} 2 & 3 \\ 4 & 7 \end{bmatrix} = \begin{bmatrix} 4 & 6 \\ 8 & 13 \end{bmatrix}$$

A matrix can be multiplied by a scalar (**scalar multiplication**) by multiplying each matrix element by the scalar (a real number), as shown here:

$$3 \cdot \begin{bmatrix} 2 & 3 \\ 4 & 7 \end{bmatrix} = \begin{bmatrix} 6 & 9 \\ 12 & 21 \end{bmatrix}$$

Matrix subtraction is the same as multiplying by the scalar of -1 and then adding:

$$\begin{bmatrix} 2 & 3 \\ 4 & 6 \end{bmatrix} - \begin{bmatrix} 1 & -2 \\ 4 & -3 \end{bmatrix} = \begin{bmatrix} 2 & 3 \\ 4 & 6 \end{bmatrix} + \begin{bmatrix} -1 & 2 \\ -4 & 3 \end{bmatrix} = \begin{bmatrix} 1 & 5 \\ 0 & 9 \end{bmatrix}$$

Entries in a matrix are referred to by using subscripts. The first subscript is the row number and the second subscript is the column number.

$$\mathbf{A} = \begin{bmatrix} 1 & 0 \\ -2 & 5 \end{bmatrix} = \begin{bmatrix} a_{11} & a_{12} \\ a_{21} & a_{22} \end{bmatrix}, \qquad a_{11} = 1, \qquad a_{12} = 0, \qquad a_{21} = -2, \qquad a_{22} = 5$$

To compute the **transpose of a matrix**, $\mathbf{A}^{\mathrm{T}}$, interchange rows and columns (make the rows columns and the columns rows).

$$\mathbf{A} = \begin{bmatrix} 1 & 3 \\ 2 & -6 \\ 4 & -1 \end{bmatrix} \qquad \mathbf{A}^{\mathrm{T}} = \begin{bmatrix} 1 & 2 & 4 \\ 3 & -6 & -1 \end{bmatrix}$$

Graphing calculators can be used to work with matrices. For example, the TI-83 has a matrix button **[MATRX]** that allows matrices to be entered and manipulated. On the TI-83 Plus, the matrices are accessed via the $[\mathbf{2^{nd}}][\mathbf{x^{-1}}]$ sequence.

 **Note:** *You should now complete Activity B in the Matrices Module.*

A linear system can be represented or written as an **augmented matrix**. The augmented matrix for the linear system from Example 3.1 is

$$2x - y = 1 \qquad \Leftrightarrow \qquad \left[\begin{array}{cc|c} 2 & -1 & 1 \\ 3 & 2 & 12 \end{array}\right]$$
$$3x + 2y = 12$$

Column 1 is the column of coefficients for the x-variables and column 2 is the column of coefficients for the y-variables. The last column is the column of numbers on the right-hand side of the equal sign. The augmented matrix uses a vertical bar in place of the equal sign. Augmented matrices allow us to represent the equations in a more compact form. We now examine matrix multiplication.

3.3   MATRIX MULTIPLICATION

To be able to multiply two matrices together, the number of columns of the first matrix must be equal to the number of rows of the second matrix. In other words, if matrix **A** has dimensions $m{\times}p$, then matrix **B** must have dimensions $p{\times}n$, in order to form the matrix product **AB**.

Note: *The new product matrix has the same number of rows as matrix **A** and the same number of columns as matrix **B**.*

To actually form the matrix product, each row of matrix **A** must be "poured down" each column of matrix **B**, producing that particular row/column entry. The "pouring down" of the rows means multiplying corresponding entries and summing the products to form each particular entry.

Example 3.5   Given the matrices below, find

a. Matrix product **AB**
b. Matrix product **BA**

$$A = \begin{bmatrix} -2 & 1 & 4 \\ 3 & 5 & 7 \end{bmatrix} \qquad B = \begin{bmatrix} 1 & 2 & 3 \\ 4 & 5 & 6 \\ 7 & 8 & 9 \end{bmatrix}$$

Solution

a. First, we must check the dimensions of the matrices to see if the matrix products are possible. To form the matrix product **AB**, the number of columns of **A**, which is 3, must equal the number of rows of **B**, which is also 3. Therefore, we can form the product matrix **AB**, which will be a 2×3 matrix, as follows:

Pour row 1 of matrix **A** down column 1 of matrix **B**, multiplying and adding as we go:

$$(-2\cdot1) + (1\cdot4) + (4\cdot7) = 30, \text{ giving us the } (ab)_{11} \text{ entry}$$

Now, pour row 1 of matrix **A** down column 2 of matrix **B**, multiplying and adding as we go:

$$(-2\cdot2) + (1\cdot5) + (4\cdot8) = 33, \text{ giving us the } (ab)_{12} \text{ entry}$$

Continuing with row 1 of matrix **A** until we run out of columns of matrix **B**, and then moving to row 2 of matrix **A**, pouring it down all columns of matrix **B**, we form the product matrix **AB**:

$$\mathbf{AB} = \begin{bmatrix} (-2\cdot1)+(1\cdot4)+(4\cdot7) & (-2\cdot2)+(1\cdot5)+(4\cdot8) & (-2\cdot3)+(1\cdot6)+(4\cdot9) \\ (3\cdot1)+(5\cdot4)+(7\cdot7) & (3\cdot2)+(5\cdot5)+(7\cdot8) & (3\cdot3)+(5\cdot6)+(7\cdot9) \end{bmatrix}$$

$$= \begin{bmatrix} 30 & 33 & 36 \\ 72 & 87 & 102 \end{bmatrix}$$

b. To try to form the matrix product **BA**, we must first check the number of columns of matrix B (3 columns) against the number of rows of matrix **A** (2 rows). Since $3 \neq 2$, the matrix product **BA** cannot be formed. ❖

Notice that $\mathbf{AB} \neq \mathbf{BA}$ in the previous example, illustrating the property that matrix multiplication is *not* commutative. In other words, the order of the matrices matters!

 **Note:** *You should now complete Activity C in the Matrices Module.*

3.4   MATRIX INVERSES

Matrices will enable us to solve linear systems of equations in more efficient ways than graphing and substitution. Two of the methods discussed here that use matrices are solving systems using **inverses** and **Gauss-Jordan elimination**.

To introduce the concept of inverses, let's consider that for any real number a (except for $a = 0$), there exists another real number a^{-1} such that the following is true:

$$a^{-1} \cdot a = a \cdot a^{-1} = 1$$

For real numbers, we may refer to this inverse as the reciprocal. If such an inverse exists for matrices, then the following must be true:

$$\mathbf{A}^{-1} \cdot \mathbf{A} = \mathbf{A} \cdot \mathbf{A}^{-1} = \mathbf{I}$$

where $\mathbf{A}$ is a square matrix and $\mathbf{I}$ is the identity matrix. $\mathbf{A}^{-1}$ and $\mathbf{A}$ are **matrix inverses**. Only square matrices can have inverses. However, just because a matrix is square doesn't necessarily mean it has an inverse. It is just a condition that must be met before we can look for the inverse. That is, the matrix $\mathbf{B}$ from Example 3.5 *might* have an inverse since it is square, but matrix $\mathbf{A}$ from that example *cannot* have an inverse since it is not square.

If a square matrix has an inverse, the matrix is called **non-singular**. If a square matrix does *not* have an inverse, the matrix is called a **singular matrix.**

Example 3.6   Find the inverse of matrix $\mathbf{A}$ (below), if it exists.

$$\mathbf{A} = \begin{bmatrix} 2 & 1 & -1 \\ -1 & -1 & 2 \\ 1 & 2 & -1 \end{bmatrix}$$

Solution

First, we look at the dimensions of $\mathbf{A}$ to determine whether or not it is square. Since $\mathbf{A}$ is 3×3 it is square and might have an inverse. One way (and perhaps the easiest way) to find the inverse of a square matrix, if it exists, is to use the graphing calculator. Enter the matrix $\mathbf{A}$ into the calculator and choose the inverse operation, $[\mathbf{x}^{-1}]$. We see that the inverse of matrix $\mathbf{A}$ is given as

$$\mathbf{A}^{-1} = \begin{bmatrix} .75 & .25 & -.25 \\ -.25 & .25 & .75 \\ .25 & .75 & .25 \end{bmatrix}$$

Note: *There is another method to calculate inverses based on the Gauss-Jordan method applied to a special form of an augmented matrix [see Chapter 3 Explorations for further references].*   ❖

To use the inverse and matrices to solve a system of linear equations, we must first decompose the system of equations into the matrix equation

$$\mathbf{A}\cdot\mathbf{X} = \mathbf{B}$$

Consider the following system of equations:

$$2x + y - z = 6$$
$$-x - y + 2z = 1$$
$$x + 2y - z = 2$$

The **coefficient matrix A** is a 3×3 square matrix consisting of the coefficients (the numbers in front of the variables) of the system.

$$\mathbf{A} = \begin{bmatrix} 2 & 1 & -1 \\ -1 & -1 & 2 \\ 1 & 2 & -1 \end{bmatrix}$$

The matrix **X** is the column matrix consisting of the variables in the system.

$$\mathbf{X} = \begin{bmatrix} x \\ y \\ z \end{bmatrix}$$

The matrix **B** is the column matrix consisting of the constants on the right side of the equal sign of the equations in the system.

$$\mathbf{B} = \begin{bmatrix} 6 \\ 1 \\ 2 \end{bmatrix}$$

So, by checking dimensions and checking that both sides are equal, we verify that our **decomposition**, $\mathbf{A} \cdot \mathbf{X} = \mathbf{B}$, works and is equivalent to the original system of equations. Now, how do we use this decomposition, matrices, and inverses to solve the system? We start with the matrix equation

$$\mathbf{A} \cdot \mathbf{X} = \mathbf{B}$$

Then, in general, if for a square matrix $\mathbf{A}$, the matrix $\mathbf{A}^{-1}$ exists, we multiply both sides of the matrix equation by $\mathbf{A}^{-1}$ *on the left-hand side* (recall that we have already established that matrix multiplication is *not* commutative, so the order of multiplication matters).

$$\mathbf{A}^{-1} \cdot \mathbf{A} \cdot \mathbf{X} = \mathbf{A}^{-1} \cdot \mathbf{B}$$

Next, we will group $\mathbf{A}$ and its inverse together

$$(\mathbf{A}^{-1} \cdot \mathbf{A}) \cdot \mathbf{X} = \mathbf{A}^{-1} \cdot \mathbf{B}$$

and use the property that a matrix multiplied by its inverse is the identity matrix $\mathbf{I}$.

$$\mathbf{I} \cdot \mathbf{X} = \mathbf{A}^{-1} \cdot \mathbf{B}$$

Now, since any matrix multiplied by the identity matrix is itself, we have solved the matrix equation for the variable matrix $\mathbf{X}$.

$$\mathbf{X} = \mathbf{A}^{-1} \cdot \mathbf{B}$$

This tells us that we can find the solution to the system, if it exists, by multiplying $\mathbf{A}^{-1}$ by the column matrix $\mathbf{B}$. Again, it cannot be stressed enough that the order of multiplication matters. It *must* be $\mathbf{A}^{-1}$ times $\mathbf{B}$.

Example 3.7 Solve the following system, using inverses.

$$2x + y - z = 6$$
$$-x - y + 2z = 1$$
$$x + 2y - z = 2$$

Solution

As we discussed above, the coefficient matrix **A** is a 3×3 square matrix consisting of the coefficients (the numbers in front) of the variables in the system.

$$\mathbf{A} = \begin{bmatrix} 2 & 1 & -1 \\ -1 & -1 & 2 \\ 1 & 2 & -1 \end{bmatrix}$$

The inverse of **A** can easily be found on the calculator as described previously, and is

$$\mathbf{A}^{-1} = \begin{bmatrix} \tfrac{3}{4} & \tfrac{1}{4} & -\tfrac{1}{4} \\ -\tfrac{1}{4} & \tfrac{1}{4} & \tfrac{3}{4} \\ \tfrac{1}{4} & \tfrac{3}{4} & \tfrac{1}{4} \end{bmatrix}$$

So to solve the given system, we must multiply $\mathbf{A}^{-1}$ by **B**:

$$\begin{bmatrix} \tfrac{3}{4} & \tfrac{1}{4} & -\tfrac{1}{4} \\ -\tfrac{1}{4} & \tfrac{1}{4} & \tfrac{3}{4} \\ \tfrac{1}{4} & \tfrac{3}{4} & \tfrac{1}{4} \end{bmatrix} \cdot \begin{bmatrix} 6 \\ 1 \\ 2 \end{bmatrix} = \begin{bmatrix} \tfrac{17}{4} \\ \tfrac{1}{4} \\ \tfrac{11}{4} \end{bmatrix}$$

Therefore, the one unique solution for this system is $x = \tfrac{17}{4}$, $y = \tfrac{1}{4}$, and $z = \tfrac{11}{4}$.   ❖

Note: *You should now complete Activity D in the Matrices Module.*

3.5   GAUSS-JORDAN ELIMINATION

Another way to solve a system of equations is to use **Gauss-Jordan elimination** with **pivoting**. In this situation, we must first decompose the system of equations into an augmented matrix.

An augmented matrix is a single matrix that contains both the coefficient matrix **A** and the matrix of constants we referred to above as **B**. To form the augmented matrix for the same system we solved using inverses, we have

$$2x + y - z = 6$$
$$-x - y + 2z = 1$$
$$x + 2y - z = 2$$

which gives us the augmented matrix

$$\left[\begin{array}{ccc|c} 2 & 1 & -1 & 6 \\ -1 & -1 & 2 & 1 \\ 1 & 2 & -1 & 2 \end{array}\right]$$

The augmented matrix and Gauss-Jordan elimination allow us to solve the system without carrying around all the baggage of the variables. The method of Gauss-Jordan elimination combines pivoting and a series of allowable **row operations** to solve the system. The three row operations we may use (with the notation at the end) to obtain an equivalent system at each step are as follows:

- Interchange any two rows ($R_i \leftrightarrow R_j$).
- Replace any row by a nonzero constant multiple of itself (for example, $3R_i \rightarrow R_i$).
- Replace any row by the sum of that row and a constant multiple of any other row (for example, $-2R_i + R_j \rightarrow R_j$).

Our goal is to apply row operations to the augmented matrix until it looks like the identity matrix augmented with a column of constants.

$$\left[\begin{array}{ccc|c} 1 & 0 & 0 & a \\ 0 & 1 & 0 & b \\ 0 & 0 & 1 & c \end{array}\right]$$

This last column of constants is the solution to our system, $x = a$, $y = b$ and $z = c$.

Pivoting occurs as we use row operations to make the diagonal entries 1, as in the matrix above, and then use the 1 (pivot off of the 1) to make zeros out of the other entries in that particular column. The Matrix Module contains the Gauss-Jordan Applet which allows the row operations to be carried out using a spreadsheet-like format.

Example 3.8   Solve the same system as in Example 3.7, but this time using Gauss-Jordan elimination and pivoting.

$$2x + y - z = 6$$
$$-x - y + 2z = 1$$
$$x + 2y - z = 2$$

Solution

The augmented matrix is formed first.

$$\begin{bmatrix} 2 & 1 & -1 & 6 \\ -1 & -1 & 2 & 1 \\ 1 & 2 & -1 & 2 \end{bmatrix}$$

The initial matrix from the applet is shown in Figure 3.11. The coefficients are entered into the cells and the system is displayed in the window below the cells.

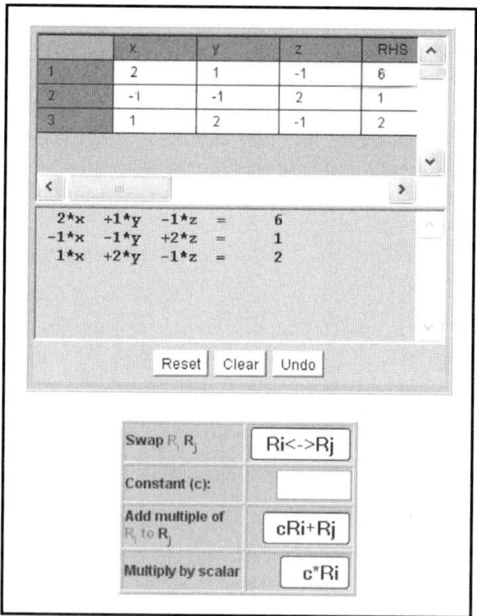

Figure 3.11   Initial Augmented Matrix from the Gauss-Jordan Applet

Step 1 We can interchange rows 1 and 3 to obtain the first pivot element (the first 1 on the diagonal). The notation for each row operation is shown in front of each row so that any changes or non-changes are clear from one matrix to the next:

$$
\begin{array}{r}
R_1 \leftrightarrow R_3 \\
R_2 \\
R_3 \leftrightarrow R_1
\end{array}
\left[
\begin{array}{ccc|c}
1 & 2 & -1 & 2 \\
-1 & -1 & 2 & 1 \\
2 & 1 & -1 & 6
\end{array}
\right]
$$

Using the applet, row 1 and row 3 are highlighted and the swap option is selected, as shown in Figure 3.12a, resulting in the second augmented matrix, shown in Figure 3.12b.

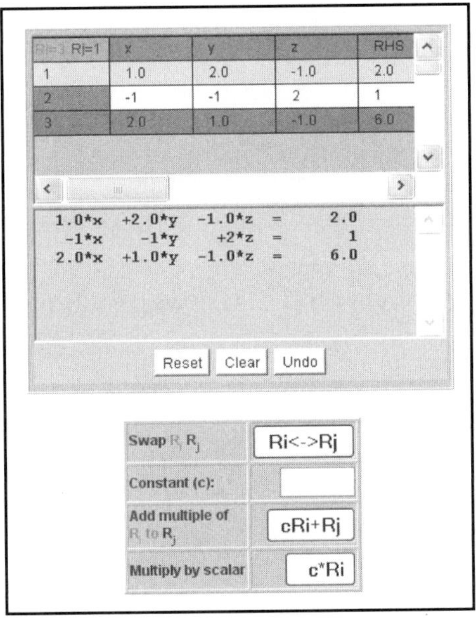

Figure 3.12a   Selecting Rows to Swap

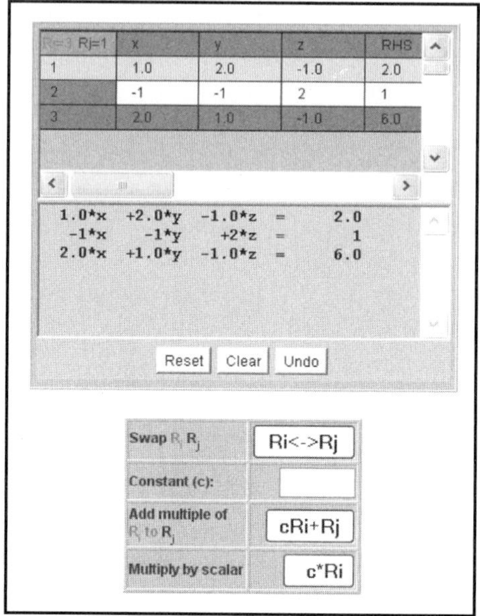

Figure 3.12b   Second Augmented Matrix from Gauss-Jordan Applet

Step 2   At this point, we use (pivot off of) the 1 we just formed to obtain zeros in the rest of the column. The row operation we use to do this is the third one, "replace any row by the sum of that row and a constant multiple of any other row."

$$\begin{matrix} R_1 \\ R_1 + R_2 \to R_2 \\ -2R_1 + R_3 \to R_3 \end{matrix} \quad \begin{bmatrix} 1 & 2 & -1 & 2 \\ 0 & 1 & 1 & 3 \\ 0 & -3 & 1 & 2 \end{bmatrix}$$

Column 1 looks good! It has a 1 as a first element and everything below it is 0.

Step 3   We move now to column 2 and according to the goal matrix, we need the second row, second column entry to be a 1. It is!

Step 4   Now, we use (pivot off of) that 1 to make the other entries, the 2 and the -3, into zeros. The row operations and equivalent matrix appear as follows.

$$\begin{array}{c} -2R_2 + R_1 \rightarrow R_1 \\ R_2 \\ 3R_2 + R_3 \rightarrow R_3 \end{array} \left[\begin{array}{ccc|c} 1 & 0 & -3 & -4 \\ 0 & 1 & 1 & 3 \\ 0 & 0 & 4 & 11 \end{array}\right]$$

Column 2 looks good! It has a 1 as the middle element and everything above and below it is 0.

Step 5   We now move to the last column, column 3, where we need to make the third row, third column entry a 1. The row operations and equivalent matrix appear below.

$$\begin{array}{c} R_1 \\ R_2 \\ \tfrac{1}{4} R_3 \rightarrow R_3 \end{array} \left[\begin{array}{ccc|c} 1 & 0 & -3 & -4 \\ 0 & 1 & 1 & 3 \\ 0 & 0 & 1 & \tfrac{11}{4} \end{array}\right]$$

Step 6   Now, we use (pivot off of) this last 1 to turn the -3 and the 1 in the last column into zeros. The row operations and equivalent matrix appear below:

$$\begin{array}{c} 3R_3 + R_1 \rightarrow R_1 \\ -R_3 + R_2 \rightarrow R_2 \\ R_3 \end{array} \left[\begin{array}{ccc|c} 1 & 0 & 0 & \tfrac{17}{4} \\ 0 & 1 & 0 & \tfrac{1}{4} \\ 0 & 0 & 1 & \tfrac{11}{4} \end{array}\right]$$

This gives us the same unique solution that we saw when we solved the system using inverses: $x = \tfrac{17}{4}$, $y = \tfrac{1}{4}$, and $z = \tfrac{11}{4}$.   ❖

Clearly, if the inverse of the coefficient matrix exists, using inverses to solve systems is much easier. However, if the inverse of the coefficient matrix does not exist, or the coefficient matrix is not square, Gauss-Jordan is a viable alternative. Let's look at an example of a system that has no solution.

Example 3.9   Solve the following system.

$$3x - 2y + 5z = 1$$
$$7x + y - 6z = -1$$
$$10x - y - z = 3$$

Solution

We start by forming the augmented matrix for the given system.

$$\begin{bmatrix} 3 & -2 & 5 & | & 1 \\ 7 & 1 & -6 & | & -1 \\ 10 & -1 & -1 & | & 3 \end{bmatrix}$$

Alternatively, we can turn to the Gauss-Jordan Applet and enter the coefficients and constants as shown in Figure 3.13.

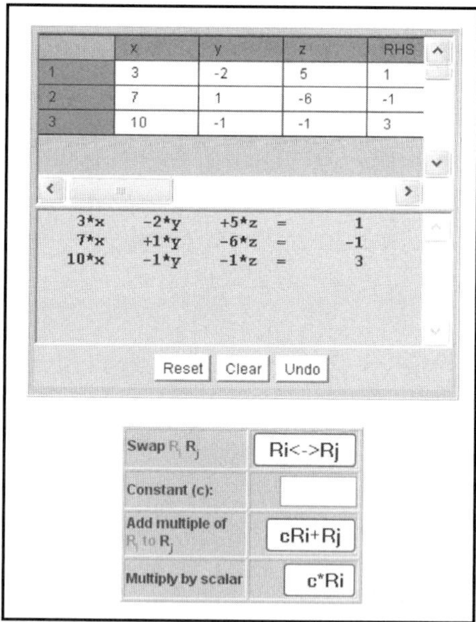

Figure 3.13   Initial Augmented Matrix from the Gauss-Jordan Applet

We now need to multiply row 1 by $\frac{1}{3}$ to form a new row 1 and get our first pivot element:

$$\frac{1}{3}R_1 \to R_1 \quad \begin{bmatrix} 1 & -\frac{2}{3} & \frac{5}{3} & \Big| & \frac{1}{3} \\ 7 & 1 & -6 & \Big| & -1 \\ 10 & -1 & -1 & \Big| & 3 \end{bmatrix} \begin{matrix} \\ R_2 \\ R_3 \end{matrix}$$

At this point, we use (pivot off of) the 1 we just formed to obtain zeros in the rest of the column. The row operation we use to do this is the third one, "replace any row by the sum of that row and a constant multiple of any other row." Using the new matrix we just formed, the next set of row operations and equivalent matrix are

$$\begin{matrix} R_1 \\ -7R_1 + R_2 \to R_2 \\ -10R_1 + R_3 \to R_3 \end{matrix} \quad \begin{bmatrix} 1 & -\frac{2}{3} & \frac{5}{3} & \Big| & \frac{1}{3} \\ 0 & \frac{17}{3} & -\frac{53}{3} & \Big| & -\frac{10}{3} \\ 0 & \frac{17}{3} & -\frac{53}{3} & \Big| & -\frac{1}{3} \end{bmatrix}$$

Column 1 looks good! It has a 1 as the first element and everything below it is 0.

We move now to column 2 and according to the goal matrix, we need the second row, second column entry to be a 1. Additionally, if we are observant, we might notice at this point that our second and third rows seem to look almost alike. Let's see what this means. Multiplying row 2 by $\frac{3}{17}$ to form the pivot element in column 2 gives us

$$\begin{matrix} R_1 \\ \frac{3}{17}R_2 \to R_2 \\ R_3 \end{matrix} \quad \begin{bmatrix} 1 & -\frac{2}{3} & \frac{5}{3} & \Big| & \frac{1}{3} \\ 0 & 1 & -\frac{53}{17} & \Big| & -\frac{10}{17} \\ 0 & \frac{17}{3} & -\frac{53}{3} & \Big| & -\frac{1}{3} \end{bmatrix}$$

Notice the nasty fractions! The graphing calculator can help a lot with the arithmetic, while we concentrate on the row operations. In particular, the program ROWOPS can be downloaded from the *Finite Math on the Web* CD under Resources/Chapter 3.

We must now "zero out" the other entries in column 2, specifically the $-\frac{2}{3}$ and $\frac{17}{3}$. The row operations and equivalent matrix are

$$\begin{matrix} \frac{2}{3}R_2 + R_1 \to R_1 \\ R_2 \\ -\frac{17}{3}R_2 + R_3 \to R_3 \end{matrix} \quad \begin{bmatrix} 1 & 0 & -\frac{7}{17} & \Big| & -\frac{1}{17} \\ 0 & 1 & -\frac{53}{17} & \Big| & -\frac{10}{17} \\ 0 & 0 & 0 & \Big| & 3 \end{bmatrix}$$

Something happened to the last row! Our third pivot element has become a zero and none of the allowable row operations will help. Additionally, the last row now says that 0 is equal to 3! This is never true, so this system has *no solution.*   ❖

Let's look at our last example and see what a system with infinitely many solutions looks like.

 **Example 3.10**   Solve the following system.

$$x + 2y + 3z = 6$$
$$4x + 5y + 6z = 15$$
$$7x + 8y + 9z = 24$$

Solution

We start by forming the augmented matrix for the given system.

$$\left[\begin{array}{ccc|c} 1 & 2 & 3 & 6 \\ 4 & 5 & 6 & 15 \\ 7 & 8 & 9 & 24 \end{array}\right]$$

Our first pivot element is already there, so we move to making zeros out of the 4 and 7 in column 1:

$$\begin{array}{c} R_1 \\ -4R_1 + R_2 \to R_2 \\ -7R_1 + R_3 \to R_3 \end{array} \left[\begin{array}{ccc|c} 1 & 2 & 3 & 6 \\ 0 & -3 & -6 & -9 \\ 0 & -6 & -12 & -18 \end{array}\right]$$

To get the second-column pivot element, we multiply row 2 by $-\frac{1}{3}$.

$$\begin{array}{c} R_1 \\ -\frac{1}{3}R_2 \to R_2 \\ R_3 \end{array} \left[\begin{array}{ccc|c} 1 & 2 & 3 & 6 \\ 0 & 1 & 2 & 3 \\ 0 & -6 & -12 & -18 \end{array}\right]$$

We must now "zero out" the other entries in column 2, the 2 and –6. The row operations and equivalent matrix are

$$
\begin{array}{r}
-2R_2 + R_1 \rightarrow R_1 \\
R_2 \\
6R_2 + R_3 \rightarrow R_3
\end{array}
\left[\begin{array}{ccc|c}
1 & 0 & -1 & 0 \\
0 & 1 & 2 & 3 \\
0 & 0 & 0 & 0
\end{array}\right]
$$

Again, something happened to the last row. Our third pivot element has become a zero and none of the allowable row operations will help. This last row now says that 0 is equal to 0! This is always true, so this system actually has two equations and three unknowns and thus *infinitely many solutions*. We can find them by writing down the equation that each of the rows in this final matrix gives us:

$$x - z = 0$$
$$y + 2z = 3$$

Now, we will solve each of these equations in terms of z, so that everything is in terms of one variable:

$$x = z$$
$$y = 3 - 2z$$

Therefore, the infinite solutions for this system look like $(z, 3 - 2z, z)$ where z can be *any* real number.   ❖

 **Note:** *You should now complete Activity E in the Matrices Module.*

 **FURTHER EXPLORATIONS**

For further explorations in the area of linear systems and matrices, please visit the URL

http://www.finitemathtutor.com/explore/chapter3/

EXERCISES

Exercise 3.1 A system of two equations and two unknowns can have how many possible solutions? Sketch an example of each type.

Find the solution to the linear systems in Exercises 3.2–3.4.

Exercise 3.2 $-5x + y = 10$
$$y = -1$$

Exercise 3.3 $-x - \tfrac{2}{3}y = 2$
$$3x + 2y = -6$$

Exercise 3.4 $x - 0.5y = 0$
$$2x + y = 6$$

Exercise 3.5   Solve the following matrix equation to find the values of a, b, c, and d.

$$2\begin{bmatrix} 3 & a \\ 0 & -1 \end{bmatrix} - \begin{bmatrix} b & 2 \\ -2 & 6 \end{bmatrix} = \begin{bmatrix} 1 & 5 \\ c & d \end{bmatrix}$$

Exercise 3.6   Find the dimensions for the following matrix products, if the product exists. Matrix **E** is a 2×4 matrix, **F** is a 4×4 matrix, **G** is a 2×2 matrix and matrix **H** is a 4×2 matrix.

a. EF	**b. FE**	**c. EG**	**d. GE**
e. HG	**f. GH**	**g. HF**	**h. FH**

Exercise 3.7   **A** is a non-singular matrix, **B** is a singular matrix, and **I** is an identity matrix. Which of the following statements are true?

a. $AA^{-1} = I$	**b. $BB^{-1} = I$**	**c. $AI = A$**	**d. $IB = B$**

Exercise 3.8 How many solutions does the linear system $\mathbf{AX} = \mathbf{B}$ have if $\mathbf{A}$ is a non-singular matrix of coefficients, $\mathbf{B}$ is a matrix of constants, and $\mathbf{X}$ is a 3×1 matrix of variables?

Use the Gauss-Jordan method or matrix inverses to solve the linear systems in Exercises 3.9 and 3.10.

Exercise 3.9
$$3x + y - 2z = 5$$
$$2x + y - z = 3$$
$$4x + 2y - z = 6$$

Exercise 3.10
$$2x + 2y - 2z = 8$$
$$x + y - z = 4$$
$$2x + y - 3z = 3$$

SAMPLE QUIZ

Question 3.1 Match the following graphs with the number of solutions of a linear system with two equations and two unknowns.

a. Two parallel lines　　　　　　　　**A.** One solution

b. Two intersecting lines　　　　　　**B.** Infinitely many solutions

c. Two lines on top of each other　　**C.** No solution

Find the solution, if any, for the systems of linear equations in Questions 3.2–3.4.

Question 3.2
$$-5x - 2y = 20$$
$$10x + 4y = 10$$

Question 3.3
$$3x - 4y = 12$$
$$y = 4$$

Question 3.4 $-4x + 2y = 8$
$$8x - 4y = -16$$

Question 3.5   Find the values of w, x, y, and z in the matrix equation below.

$$\begin{bmatrix} w & x \\ y & z \end{bmatrix} + 3\begin{bmatrix} -1 & 0 \\ 4 & 2 \end{bmatrix} = \begin{bmatrix} 5 & 3 \\ 1 & -3 \end{bmatrix}$$

Question 3.6   Given that matrix **E** is a 1×5 matrix, **F** is a 5×1 matrix, **G** is a 5×5 matrix, and **H** is a 1×1 matrix, find the dimensions of the following products, if they exist.

a. EF	**b. FE**	**c. EG**	**d. GE**
e. HG	**f. GH**	**g. HF**	**h. FH**

Question 3.7   Which of the following statements are false?

a. Every square matrix has an inverse.
b. Some non-square matrices have inverses.
c. Multiplying a matrix by the identity matrix leaves the matrix unchanged.
d. The product of a matrix and its inverse is the identity matrix.

Question 3.8   Which of the following conditions must be met before a linear system can be solved using matrix inverses?

a. There must be the same number of equations and variables.
b. There must be more equations than variables.
c. There must be fewer equations than variables.
d. The system must have infinitely many solutions.
e. The system must have exactly one solution.
f. The system must have no solution.

Use matrix inverses or the Gauss-Jordan method to solve the systems in Questions 3.9 and 3.10.

Question 3.9
$$-x - 4y + z = 6$$
$$x + 3y - z = -5$$
$$2x + 6y - z = -8$$

Question 3.10
$$3x - y + 3.5z = 7$$
$$x - y + 2.5z = 3$$
$$-2x + y - 3z = -5$$

Linear Programming

4

In real-world problems, solutions are often governed by "constraints." For example, if we manufacture computers, the number of items produced must be an integer – we cannot manufacture $\frac{1}{13}$ of a computer! Clearly, we also cannot produce negative numbers of items. These are examples of constraints. The first restricts the "admissible" solutions to the set of integers. The second constraint is an inequality, restricting the "admissible" solutions to positive integers. Other constraints include the manufacturing capacity of plants, number of hours of overtime allowed by law, and shipping deadlines.

Often, problems are cast in the form of minimization or maximization problems. An example is the classic **traveling salesman problem**. Given that a salesman must visit 10 cities, find the minimum distance he must travel in order to visit each of the cities at least once. Clearly, this is a practical problem, and is a good example of what are called scheduling problems.

If we are trying to find the minimum or maximum of a linear function, subject to linear inequalities (constraints), we are solving a **linear programming problem**. In this chapter we discuss some minimization problems, but focus on maximization problems.

4.1   LINEAR PROGRAMMING PROBLEMS

Suppose one has a function of a single variable, say $f(x)$, and wants to find the maximum (or minimum) of the function subject to two constraints, $x \geq b$ and $x \leq a$. If the function is linear, as shown in Figure 4.1, one can easily see that the minimum and maximum must occur at the endpoints. If the function is increasing, the minimum will be the left endpoint, and the maximum will be the right endpoint. If the function is decreasing, the minimum will be at the right endpoint, and the maximum will be at the left endpoint.

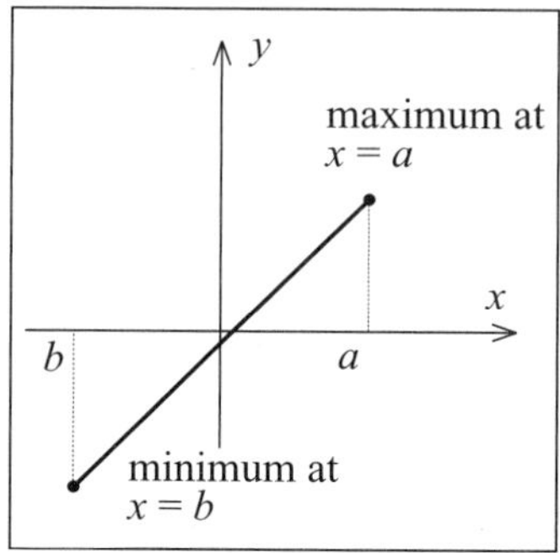

Figure 4.1   Minimum and Maximum of an Increasing Linear Function

If the function is not linear, then the minimum or maximum could be at some other point in the interior. However, here we must distinguish between a relative minimum and an absolute minimum. A **relative minimum** is a value less than values that are nearby, and an **absolute minimum** is the smallest value the function takes on. Similarly, a **relative maximum** is a value greater than values nearby, and an **absolute maximum** is the largest value the function can achieve. In Figure 4.2 we illustrate the relative maximum and minimum of a non-linear function. As shown in Figure 4.2, it is not easy to determine the exact location of the minimum and maximum of a non-linear function.

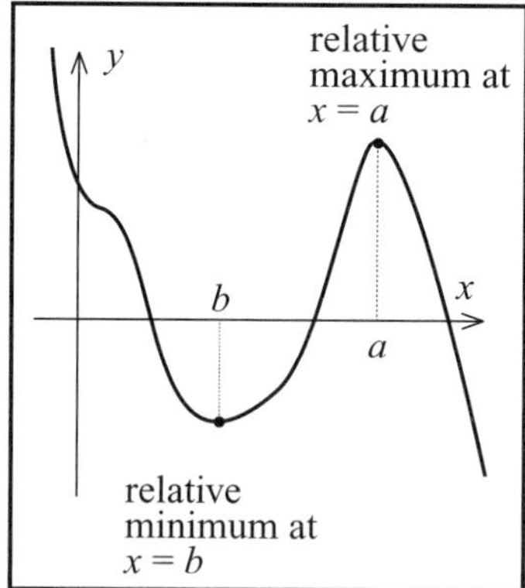

Figure 4.2   Relative Minimum and Maximum of a Non-Linear Function

Extending this case to a function of two variables, x and y, we can envision even more complex situations. The simplest case is a linear function of x and y of the form

$$f(x, y) = Ax + By + C$$

where there are one or more linear inequalities (constraints) of the form

$$ax + by \geq c$$

The solution to a linear inequality of this type is a half-plane. The line $ax + by = c$ divides the xy-plane into an upper and lower half-plane. One half-plane is true for the inequality and one is false. The half-plane that is false is shaded and the solution is the unshaded region called the **feasible region**. We will label the feasible region with an S.

When there is a system of inequalities, the inequalities are examined one at a time and the region of the xy-plane that has not been shaded is the solution to the system. The inequalities $\{x \geq 0, x \leq 1, y \geq 0, y \leq 1\}$ describe a unit square. The inequality $y \leq x$ describes all the points below the line $y = x$ (half-plane). As an example of a non-linear inequality constraint, the set of points $\{x^2 + y^2 \leq 1\}$ describes all points that lie inside (or on) a unit circle. These feasible regions are shown in Figure 4.3.

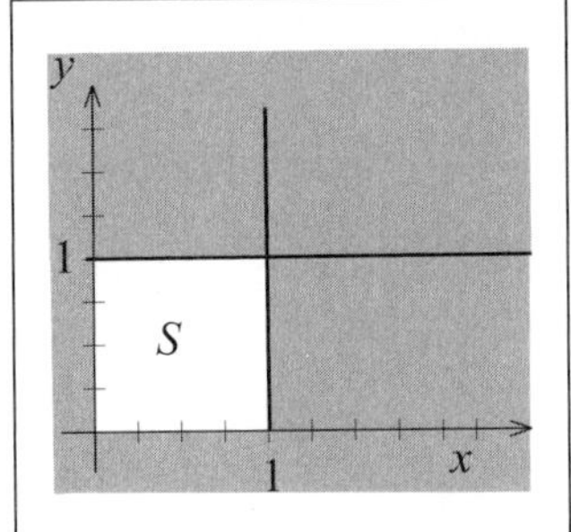
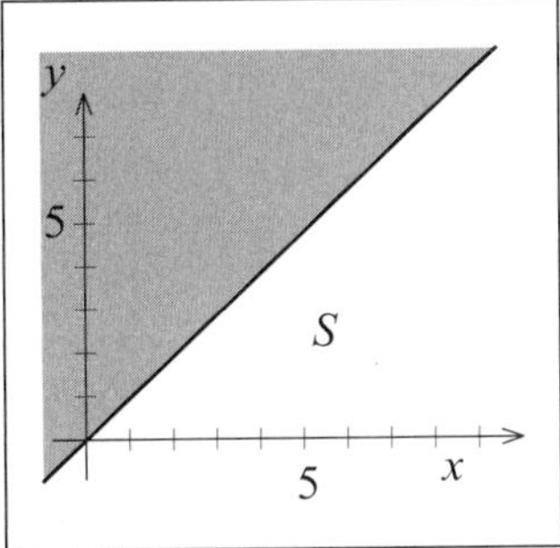
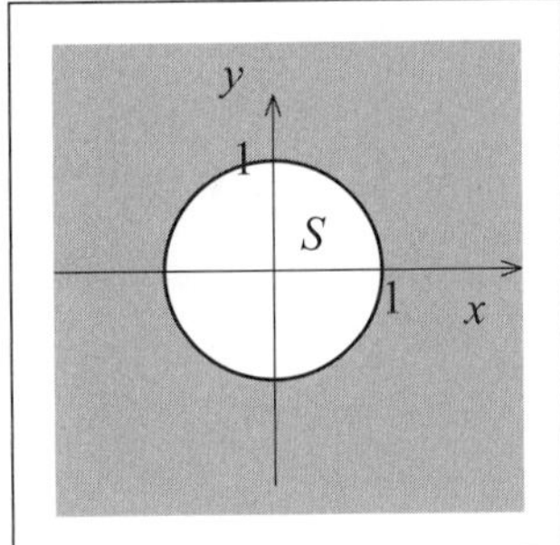

Figure 4.3   Some Feasible Regions

There are many ways to try to solve optimization problems. We describe two procedures – the *method of corners* and the *simplex algorithm*.

4.2   METHOD OF CORNERS

The **method of corners** is the simplest optimization procedure in the case of a linear function of two variables subject to linear constraints. When we have a set of more than one linear constraint, the set of points that satisfy all the constraints usually forms the interior of a **polygon**. Furthermore, the set of points is usually **closed** and **convex**. The feasible region can be graphed traditionally, as shown in Figure 4.4, or using the Method of Corners Applet, as shown in Figure 4.5.

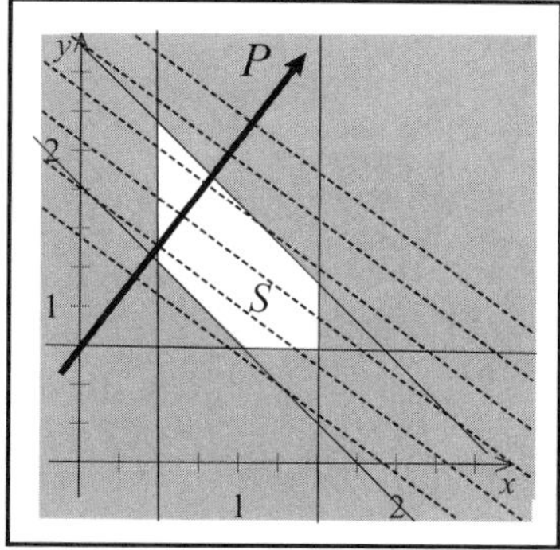

Figure 4.4   Feasible Set in Two Dimensions, Traditional Graphing Method.

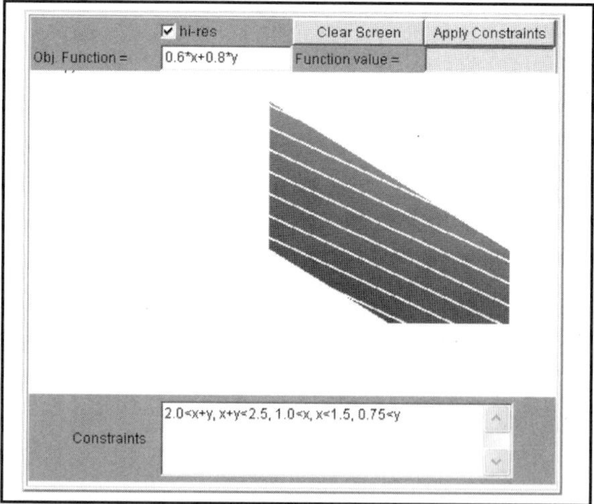

Figure 4.5   Feasible Set in Two Dimensions, Applet Method

Our problem is to maximize (or minimize) a function of two variables, called the **objective function**

$$P = Ax + By + C$$

over such a region. Curves along which P is constant are straight lines, and cover the region completely. The direction of increase (or decrease) in P is perpendicular to these lines. When using the traditional graphing, we show where the objective function is constant with dashed lines. Perpendicular to these dashed lines is the direction the objective function is increasing. This is shown in Figure 4.4 as a dark arrow. An advantage to the applet method is that the shading lightens as the value of the objective function increases. It is clear that P has its maximum and minimum on the boundary of the constraint set, in fact at one of the corners!

The method of corners reduces to the following procedure:

- Find the set of corners (or vertices) of the feasible region where two constraint equations intersect.
- Create a table of values for P at each of the corners.
- Find the largest and smallest values of P. These will satisfy the optimization problem if the feasible region is bounded (a closed polygon).

Example 4.1   Graph the feasible region for the following system of linear inequalities:

$$x + y \leq 9$$
$$2x + 5y \leq 30$$
$$x \geq 0, \ y \geq 0$$

Solution

Graph the line $x + y = 9$. Use the origin to test which half-plane is true for the original inequality, $0 + 0 \leq 9$. This statement is true, so the half-plane containing the origin is the solution to the inequality $x + y \leq 9$. The false region for this inequality is then shaded. Next the line $2x + 5y = 30$ is graphed and the origin used as a test point. The region containing the origin is found to be true and is therefore the solution to this inequality. The statements $x \geq 0$ and $y \geq 0$ restrict the solution to the first quadrant. The feasible region is shown in Figure 4.6 as a traditional graph and in Figure 4.7 as an applet graph.

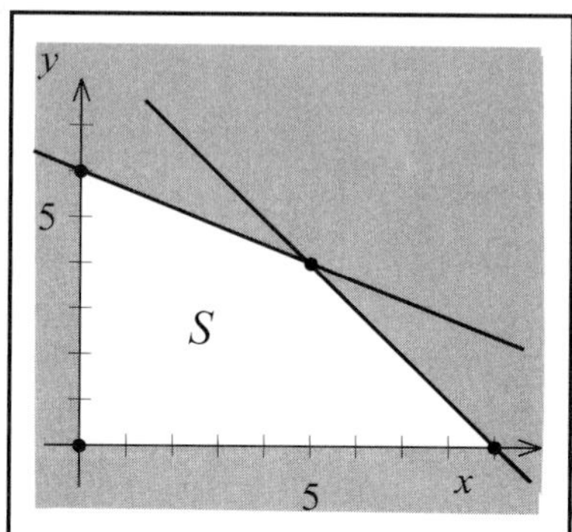

Figure 4.6   Feasible Region Using a Traditional Graph

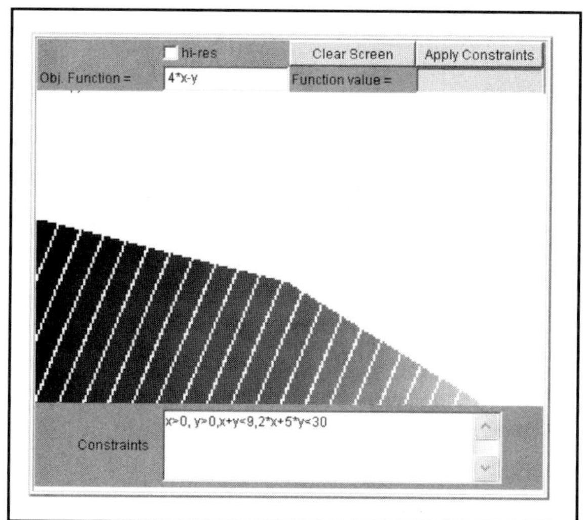

Figure 4.7   Feasible Region Using the Method of Corners Applet   ❖

 **Example 4.2**   Find the exact coordinates of the corners of the feasible region of Example 4.1.

Solution

The intersection of the line $x + y = 9$ and the line $3x + 5y = 30$ is $(5, 4)$. The remaining corners are intercepts and can be found at $(0, 0)$, $(0, 6)$, and $(9, 0)$.   ❖

Example 4.3   What are the maximum and minimum values of the objective function $P = 4x - y$ in the feasible region of Example 4.1?

Solution

Arrange the vertices in a table and evaluate the objective function at each vertex. The vertex with the largest value is the maximum and the vertex with the smallest value is the minimum.

Vertex	$P = 4x - y$	
(0, 0)	0	
(0, 6)	–6	Minimum
(5, 4)	16	
(9, 0)	36	Maximum

The maximum value is 36 at (9, 0) and the minimum value is –6 at (0, 6).   

 **Note:**   *You should now complete Activities A, B, and C in the Linear Programming Module.*

For three independent variables (x, y, z), the feasible set is a polyhedron in three dimensions. The feasible region can be visualized in three dimensions by using the Simplex Applet (Figure 4.8).

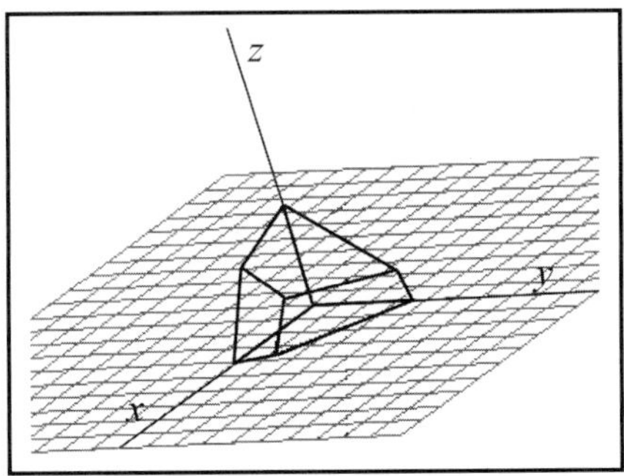

Figure 4.8   Example of a Feasible Set in Three Dimensions

Unfortunately, in more than two or three dimensions one cannot visualize the constraint set, and the cost of calculating all the possible vertices becomes very expensive. This is why an alternative non-graphical procedure was created – to solve linear constraint problems in an arbitrary number of variables.

4.3   SIMPLEX METHOD

The idea underlying the **simplex algorithm** is the same as the method of corners – the maximum of a linear function occurs on the boundary of the constraint set. The contribution of the simplex algorithm lies in the particular way that the vertices are selected. Given a starting point, each vertex is selected in such a way that the objective function, P, is increased (or decreased) at each stage.

The starting point of the simplex algorithm is to convert all the **inequality constraints** to equalities. We do this by introducing **slack variables**. For example, if $x + y \leq 10$, then $10 - x - y \geq 0$, and we can introduce a new variable called u where $u = 10 - x - y$ and $u \geq 0$. Therefore the original inequality can be written as $x + y + u = 10$. Each inequality introduces another variable by this procedure. This leads to an **over-determined system of equations** after the addition of the slack variables, which has more unknowns than equalities, and which has a non-unique solution. The particular solution we desire maximizes the objective function.

The beauty of the simplex algorithm is that it uses the techniques of pivoting to find the appropriate solution of our over-determined system of equations. Chosing the pivot elements according to a certain definite procedure does this. To motivate this procedure used to determine the pivot element, consider a particular objective function we wish to maximize:

$$P = x - 2y + 3z$$

If all the variables are positive, it is clear that at the maximum, $y = 0$! Why is this? If $y > 0$, the value of P actually decreases. We see that as long as we wish to maximize the objective function P, we set the values of any independent variable to zero if it occurs with a negative coefficient. Another way of phrasing this is to say that any variable with a positive coefficient in

$$-x + 2y - 3z + P = 0$$

must vanish. Also, given a choice between changing the value of x one unit, and z one unit, we see that z makes the most difference. This leads us to choose the variables with the largest negative coefficients. A more detailed explanation of the simplex algorithm may be found in your textbook.

The Simplex Algorithm

1. Find the column with the *most negative coefficient* in the last row; this is called the pivot column. This column represents the variable with the most rapid change.
2. Find the row with the *smallest positive ratio* of the right-hand side to the pivot column coefficient; this is called the pivot row. This row determines the maximum change that doesn't violate the constraints.
3. The pivot element is the intersection of the pivot row and pivot column.
4. Using elementary row operations, reduce the elements above and below the pivot element. This introduces new basic variables and slack variables.
5. Repeat steps 1 to 4 until there are *no negative elements* in the last row. The maximum value has been achieved.
6. Set the values of the variables corresponding to positive coefficients in the last row to zero.
7. Solve for the remaining variables, and substitute into the objective function.

Example 4.4   Set up and solve the following linear programming problem:

$$\text{Maximize} \quad P = 2x + 3y + 8z$$
$$\text{Subject to:} \quad 5x + y + 3z \le 15$$
$$x + 4y + 6z \le 18$$
$$4x + 8y + z \le 16$$
$$x \ge 0,\ y \ge 0,\ z \ge 0$$

Solution

Begin by converting the inequalities to equations by introducing slack variables and rewriting the objective function.

$$5x + y + 3z + u = 15$$
$$x + 4y + 6z + v = 18$$
$$4x + 8y + z + w = 16$$
$$-2x - 3y - 8z + P = 0$$

Next put the linear system into an augmented matrix (see Chapter 3 for details). In a simplex tableau an additional line is drawn between the linear system coefficients and the objective function coefficients. This is a reminder that the objective function coefficients guide the selection of the pivot row, but cannot be pivot elements since they are not corners of the feasible region.

The first pivot column is column 3 since -8 is the most negative entry in the bottom row. The pivot row is row 2 since the ratio of its right-hand side to its pivot column entry is the smallest. The element we pivot on is the 6, as shown.

$$\left[\begin{array}{ccccccc|c}
5 & 1 & 3 & 1 & 0 & 0 & 0 & 15 \\
1 & 4 & \langle 6 \rangle & 0 & 1 & 0 & 0 & 18 \\
4 & 8 & 1 & 0 & 0 & 1 & 0 & 16 \\
\hline
-2 & -3 & -8 & 0 & 0 & 0 & 1 & 0
\end{array}\right]
\begin{array}{l}
15/3 = 5 \\
18/6 = 3 \\
16/1 = 16 \\
\\
\end{array} \Rightarrow$$

After the pivot, the most negative entry in the bottom row is $-\frac{2}{3}$ so the pivot column is column 1. The smallest ratio is in row 1. After the second pivot there are no negative numbers in the bottom row, so the maximum value has been reached.

$$\left[\begin{array}{ccccccc|c} \langle 9/2 \rangle & -1 & 0 & 1 & -1/2 & 0 & 0 & 6 \\ 1/6 & 2/3 & 1 & 0 & 1/6 & 0 & 0 & 3 \\ 23/6 & 22/3 & 0 & 0 & -1/6 & 1 & 0 & 13 \\ \hline -2/3 & 7/2 & 0 & 0 & 4/3 & 0 & 1 & 24 \end{array}\right] \quad \begin{array}{l} 6/(9/2) = 12/9 \approx 1.3 \\ 3/(1/6) = 18 \\ 13/(23/6) = 78/23 \approx 3.4 \end{array} \quad \Rightarrow$$

$$\left[\begin{array}{ccccccc|c} 1 & -2/9 & 0 & 2/9 & -1/9 & 0 & 0 & 4/3 \\ 0 & 19/27 & 1 & -1/27 & 5/27 & 0 & 0 & 25/9 \\ 0 & 221/27 & 0 & -23/27 & 7/27 & 1 & 0 & 71/9 \\ \hline 0 & 59/27 & 0 & 4/27 & 34 & 0 & 1 & 224/9 \end{array}\right]$$

To read the final tableau, set the value of variables on which you *did not* pivot to zero and read the value of the variable on which you did pivot from the last column:

$$x = \frac{4}{3}, \quad y = 0, \quad z = \frac{25}{9}, \quad u = 0, \quad v = 0, \quad w = \frac{71}{9}, \quad P = \frac{224}{9}$$

Alternatively, the Simplex Applet can be used to set up, graph, and solve the problem. The screenshot in Figure 4.9 shows the applet being used to start the simplex problem. Figure 4.10 shows the final tableau.

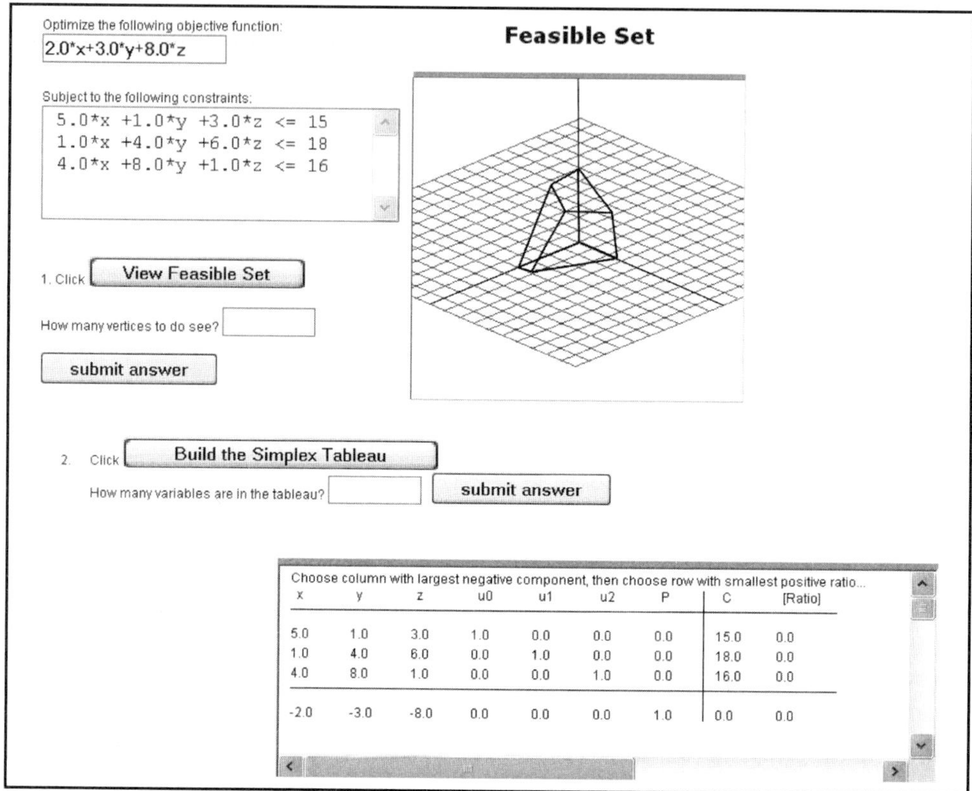

x	y	z	u0	u1	u2	P	C	[Ratio]
5.0	1.0	3.0	1.0	0.0	0.0	0.0	15.0	0.0
1.0	4.0	6.0	0.0	1.0	0.0	0.0	18.0	0.0
4.0	8.0	1.0	0.0	0.0	1.0	0.0	16.0	0.0
-2.0	-3.0	-8.0	0.0	0.0	0.0	1.0	0.0	0.0

Figure 4.9   The Simplex Applet

x	y	z	u0	u1	u2	P	C	[Ratio]
1.0	-0.222	0.0	0.2222	-0.111	0.0	0.0	1.3333	0.2962
0.0	0.7037	1.0	-0.037	0.1851	0.0	0.0	2.7777	17.950
0.0	8.1851	0.0	-0.851	0.2592	1.0	0.0	7.8888	2.2555
0.0	2.1851	0.0	0.1481	1.2592	0.0	1.0	24.888	4.1975

Figure 4.10   Final Simplex Tableau from Applet

 **Note:** *You should now complete Activities B – E in the Linear Programming Module.*

FURTHER EXPLORATIONS

For further explorations in the area of linear programming, please visit the URL

http://www.finitemathtutor.com/explore/chapter4/

EXERCISES

Exercise 4.1   Graph the feasible region for the following system of linear inequalities:

$$x + 3y \le 9$$
$$5x + 2y \le 10$$
$$x \ge 0, y \ge 0$$

Exercise 4.2   Find the exact coordinates of all corners of the feasible region found in Exercise 4.1.

Exercise 4.3   What are the maximum and minimum values of the objective function $P = 3x - 2y$ in the feasible region from Exercise 4.1?

Exercise 4.4   Set up the following linear programming problem: The Summer Day chair company makes two kinds of chairs, rocking and lounging. A rocking chair uses 16 units of wood and 2 hours of assembly time and sells for $100. A lounging chair uses 10 units of wood and 2.5 hours of assembly time and sells for $160. Every day the company has 240 units of wood and 40 hours of assembly time available. How many of each kind of chair should be made to maximize revenue?

Exercise 4.5   Find the solution to Exercise 4.4.

Exercise 4.6   Set up the initial simplex tableau and circle the first pivot element for the following linear programming problem:

$$\text{Maximize} \quad P = 4x + y + 5z$$
$$\text{Subject to:} \quad x + 3y + 5z \leq 15$$
$$6x + y + 2z \leq 18$$
$$4x + 2y + z \leq 16$$
$$x \geq 0, y \geq 0, z \geq 0$$

Exercises 4.7 and 4.8 use the linear programming problem from Exercise 4.6.

Exercise 4.7   What is the value of the objective function at the maximum?

Exercise 4.8   Find the values of all the variables when the objective function is maximized.

Exercise 4.9   Set up the initial simplex tableau and circle the first pivot element for the following linear programming problem:

$$\text{Maximize} \quad P = x + 2y + 4z$$
$$\text{Subject to:} \quad 2x - y + 3z \leq 24$$
$$x + 5y \leq 15$$
$$3x + 2z \leq 18$$
$$x \geq 0, y \geq 0, z \geq 0$$

Exercise 4.10   What is the maximum value of P in Exercise 4.9?

SAMPLE QUIZ

Use the following linear programming problem to answer Questions 4.1–4.5.

$$\text{Maximize} \quad P = 3x - 2y$$
$$\text{Subject to:} \quad x + y \leq 10$$
$$x \geq 0$$
$$y \geq 0$$

Question 4.1 How many corners does the feasible region have?

Question 4.2 Graph the objective function $P = 3x - 2y$ for $P = -12$, 0, and 12. In what direction is the objective function increasing?

Question 4.3 Find the coordinates of all corners of the feasible region.

Question 4.4 At which corner is the objective function the largest?

Question 4.5 The objective function is maximized at the intersection of which two lines?

Use the following linear programming problem to answer Questions 4.6–4.10.

$$\begin{aligned}
\text{Maximize} \quad & P = 20x + 12y + 18z \\
\text{Subject to:} \quad & 3x + y + 2z \leq 9 \\
& 2x + 3y + z \leq 8 \\
& x + 2y + 3z \leq 7 \\
& x \geq 0,\ y \geq 0,\ z \leq 0
\end{aligned}$$

Question 4.6 Write the initial simplex tableau.

Question 4.7 How many variables are in the simplex tableau?

Question 4.8 What is the first pivot element? Pivot on this element. What is the value of the objective function after pivoting?

Question 4.9 Find the next pivot element and perform the pivot. What is the value of the objective function after pivoting?

Question 4.10 Pivot on the final pivot element. What is the maximum value of the objective function?

Sets

5

One of the most fundamental concepts in mathematics is that of a **collection** or *group* of objects. A group of objects is called a **set**. Members of a set often share similar properties or characteristics, such as a set of integers, or a set of geometric shapes, or a set of colors. If we can count the number of objects in a set, we say the set is **finite**. The number of elements in a set is called its **cardinality**.

5.1   REVIEW OF SETS

Sets can be denoted in many ways. One common method is to list each of the elements in the set between a pair of braces. This is known as **roster notation**. However, sometimes there are too many elements to list and it is more convenient to use **set-builder notation,** where the properties needed for membership in the set are given.

Example 5.1   Let A be the set of all integers from 0 to 100. Write A using roster notation and using set-builder notation.

Solution
Roster notation: $A = \{0, 1, 2, 3, 4, \ldots, 100\}$
Set-builder notation: $A = \{x \mid x$ is an integer from 0 to 100$\}$

In words, the set-builder notation reads: "A is the set of all x such that x is an integer from 0 to 100." ❖

Several key ideas are central to discussing sets. The first is **set membership**. An object is either in a set, or it is not. If an object x is in a set A we write

$$x \in A$$

The symbol $\in$ means "is an element of." If x is not in the set A, we write

$$x \notin A$$

If every element of a set A is also an element of a set B, then A is called a **subset** of B and is written as

$$A \subseteq B$$

Moreover, if $A \subseteq B$, but $A \neq B$ (they do not have exactly the same elements), then A is called a **proper subset** of B and is written as

$$A \subset B$$

Example 5.2   Let $A = \{0, 1, 2, 3, 4, 5\}$, $B = \{1, 2, 3\}$, $C = \{0, 2, 4\}$, and $D = \{3, 2, 1\}$. Which of the following statements are true?

a. $B = D$ **b.** $C \subseteq A$ **c.** $B \subseteq D$ **d.** $D \subset B$ **e.** $A \subset B$

Solution

The statements **a**, **b**, and **c** are true. The statement **a** is true because B and D contain the exact same elements. The statement **b** is true because each element in C is also in A. Notice that $C \subset A$ would also be a true statement since $C \subseteq A$ but $C \neq A$. The statement **c** is true because each element in B is also in D. The statement **d** is false because even though every element in D is in B, $B = D$ so D is not a proper subset of B. Finally, the statement **f** is false since there are elements in A (0, 4, and 5) that are not in B.   ❖

There are two special sets, the **empty set** and the **universal set**. The empty set is the set which contains nothing (no elements) and is denoted by either of the following:

$$\varnothing = \{\}$$

The universal set, U, contains everything of interest.

There is a deep relationship between **Boolean logic** and *set membership*. For example, the **union** of two sets A and B is written as

$$A \cup B$$

This is the set of objects that are either only in A or only in B *or in both*. (This is known as the **inclusive or** in Boolean logic. The **exclusive or** is only **A** or only **B** but *not* both.) We can write the union in terms of a rule

$$A \cup B = \{x \mid x \in A \text{ or } x \in B\}$$

The **intersection** of two sets A and B is the set of objects that are in A *and B*.

$$A \cap B = \{x \mid x \in A \text{ and } x \in B\}$$

The **complement** of the set A is the set of objects that are in the universal set, but are *not* in A.

$$A^c = \{x \mid x \in U \text{ and } x \notin A\}$$

Our two special sets are related by the following identities:

$$\varnothing = U^c, \quad A \cup A^c = U, \quad A \cap A^c = \varnothing$$

Example 5.3   Let $U = \{1, 2, 3, \ldots, 19\}$, and let A be the set of all positive integers less than 10, and B the set of all positive even integers less than 20. Find $A \cap B$ and $A^c \cap B$.

Solution

In roster notation we have

$$A = \{1, 2, 3, 4, 5, 6, 7, 8, 9\} \text{ and } B = \{2, 4, 6, 8, 10, 12, 14, 16, 18\}.$$

so find $A \cap B = \{1, 2, 3, 4, 5, 6, 7, 8, 9\} \cap \{2, 4, 6, 8, 10, 12, 14, 16, 18\} = \{2, 4, 6, 8\}$. We also compute that $A^c = \{10, 11, 12, 13, 14, 15, 16, 17, 18, 19\}$, therefore $A^c \cap B = \{10, 12, 14, 16, 18\}$. Notice that $(A \cap B) \cup (A^c \cap B) = B$.   ❖

5.2   VENN DIAGRAMS

When we have just two sets to consider, it is very helpful to use the following diagrams.

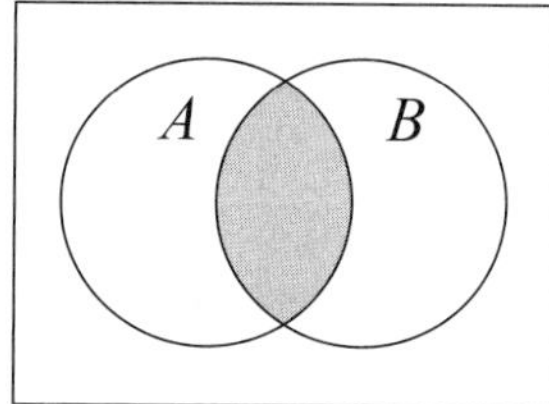

Figure 5.1   Intersection of Sets A and B

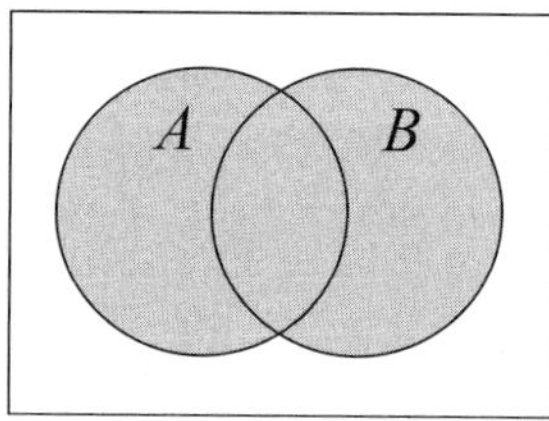

Figure 5.2   Union of Sets A and B

In Figure 5.1, the shaded region is the set of elements that are in both *A and B* at the same time, and therefore represents the *intersection* of the sets *A* and *B*. In Figure 5.2, the shaded region is the set of elements that are in *A or B* (*or both*), and therefore represents the *union* of the sets *A* and *B*. These diagrams are special cases of **Venn diagrams**. Figures 5.3a and 5.3b show screenshots of the main Set Building Applet.

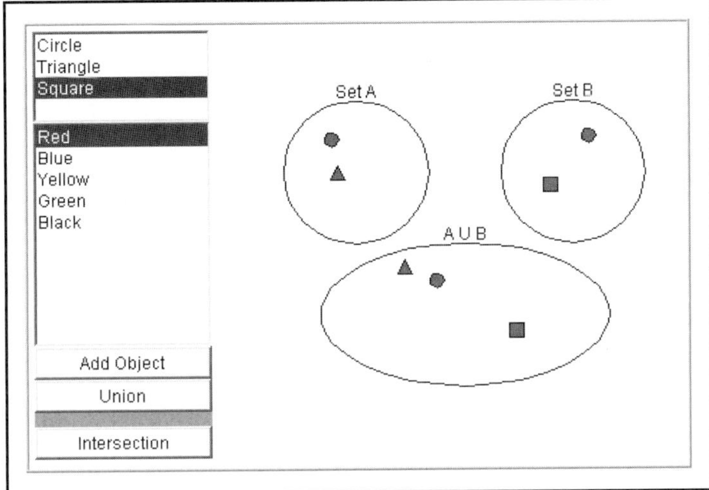

Figure 5.3a   Set Building Applet Showing A ∪ B

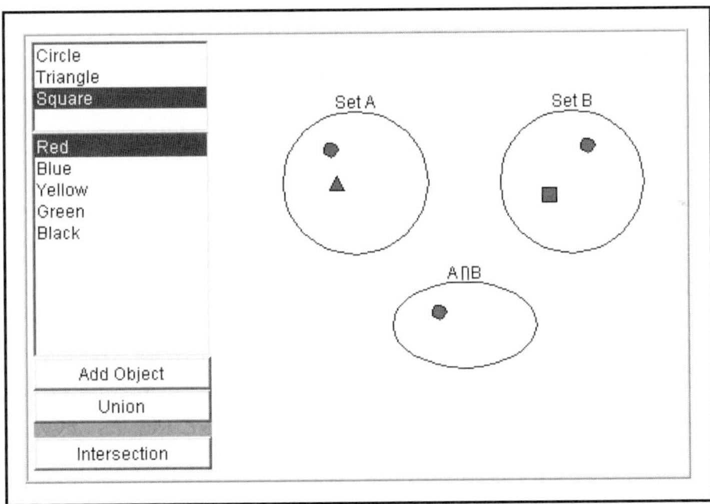

Figure 5.3b   Set Building Applet Showing A ∩ B

In this applet, one can select three different geometrical shapes and five colors (for a total of 15 different combinations). Pressing "Add Object" adds a shape on the screen, and one can drag it to a position inside Set *A* or inside Set *B*. After selecting and placing a number of different shapes, the user presses the "Union" or "Intersection" button to see the result of $A \cup B$ or $A \cap B$.

 **Note:** *You should now complete Activities A and B in the Sets Module.*

If we have three sets, we have many more possibilities. Figure 5.4 shows a Venn Diagram Applet for three sets. There are eight distinct regions. (Region VIII lies outside all three sets.)

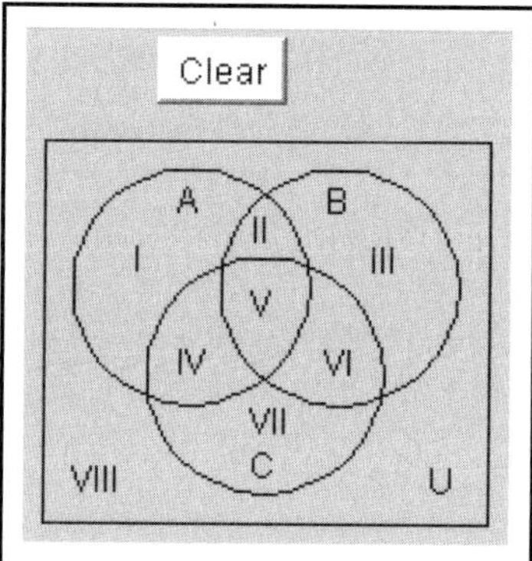

Figure 5.4   Venn Diagram for Three Sets

In this applet, clicking the mouse inside one of the labeled regions (such as Region IV) highlights it (Figure 5.5). The "Clear" button will remove all highlighting.

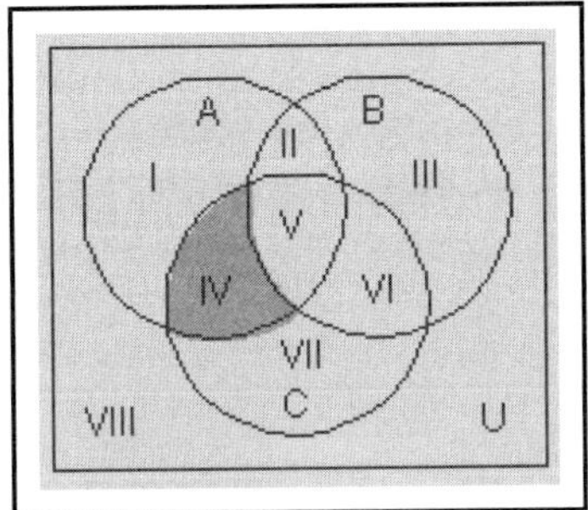

Figure 5.5 Selecting a Region

Regions {II, V} together give the intersection of A and B. The regions {II, III, IV, V, VI, VII} give the union of B and C. Any combination of unions, intersections, and complements can be represented by an appropriate combination of regions I–VIII. Figure 5.6 illustrates a more complicated set.

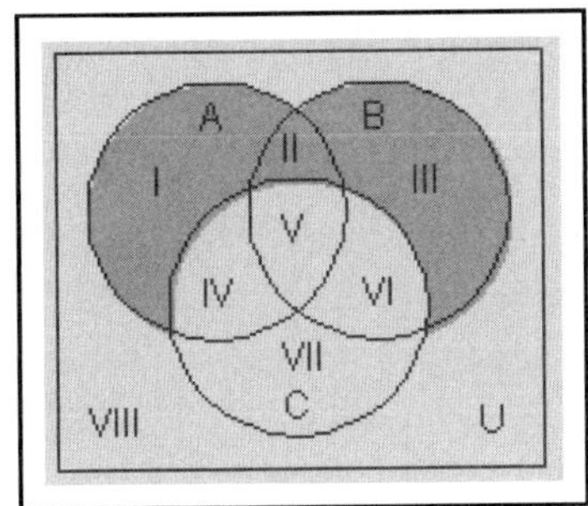

Figure 5.6 $(A \cup B) \cap C^c$

 **Note:** *You should now complete Activity C in the Sets Module.*

5.3   COUNTING ELEMENTS IN SETS

Suppose we know the number of elements in two sets A and B. Can we calculate the number of elements in $A \cup B$? Let $n(A)$ be the number of elements in set A, and $n(B)$ be the number of elements in set B. If A and B do not overlap (their intersection is empty), then clearly $n(A \cup B) = n(A) + n(B)$, as shown in Figure 5.7.

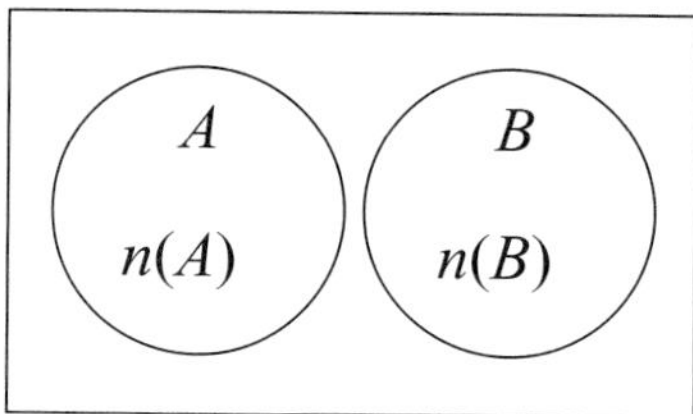

Figure 5.7   Non-Overlapping Sets

 **Example 5.4**   Let $A = \{a, e, i, o, u\}$ and $B = \{b, c, f\}$. Find $n(A \cup B)$.

Solution

$A \cup B = \{a, b, c, e, f, i, o, u\}$, so $n(A \cup B) = 8$ by counting the elements. Also, $n(A) = 5$, $n(B) = 3$ and since $A \cap B = \emptyset$, $n(A \cap B) = 0$. Then $n(A \cup B) = n(A) + n(B) = 5 + 3 = 8$. ❖

What happens if the two sets overlap, and the intersection is not empty (Figure 5.8)?

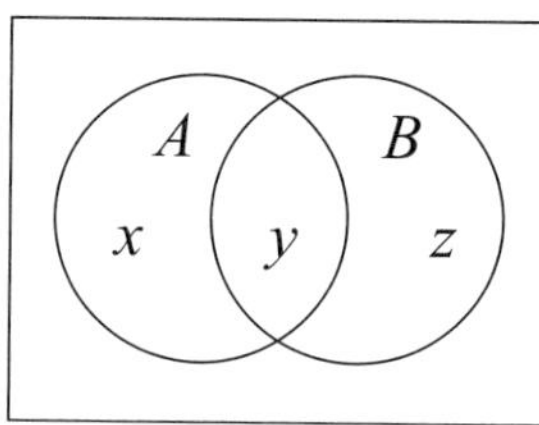

Figure 5.8   Overlapping Sets

It is clear that $n(A) = x + y$, $n(B) = y + z$, and $n(A \cap B) = y$, so we have the following relationship, known as the *Union Rule*:

$$n(A\cup B) = x + y + z = (x + y) + (y + z) - y$$
$$= n(A) + n(B) - n(A\cap B)$$

Example 5.5　Let $A = \{s, t, u, v, w, x, y, z\}$ and $B = \{a, x, y, z\}$. Find $n(A \cup B)$.

Solution
Directly $A \cup B = \{a, s, t, u, v, w, x, y, z\}$, therefore $n(A \cup B) = 9$. Furthermore, $A \cap B = \{x, y, z\}$, so $n(A) = 8$, $n(B) = 4$, and $n(A \cap B) = 3$. Thus, $n(A \cup B) = n(A) + n(B) - n(A \cap B) = 8 + 4 - 3 = 9$.　❖

The fundamental counting law for two sets can be extended to three sets $\{A, B, C\}$ by means of the formula

$$n(A\cup B\cup C) = n(A) + n(B) + n(C)$$
$$-n(A\cap B) - n(B\cap C) - n(A\cap C)$$
$$+n(A\cap B\cap C)$$

Example 5.6　A survey of high school students is conducted to find out their pizza topping preferences: pepperoni, sausage, or olives. The following information is collected:

- 190 students like pepperoni
- 40 students like sausage and olives
- 130 students like olives
- 150 students like pepperoni and sausage
- 20 students like olives and pepperoni
- 200 students like sausage
- 5 students like none of these toppings
- 5 students like all three of these toppings

Fill in a Venn diagram that represents the collected data.

Solution

First, let

$U = \{x \mid x$ is a student surveyed$\}$
$P = \{x \in U \mid x$ is a student who likes pepperoni$\}$
$S = \{x \in U \mid x$ is a student who likes sausage$\}$
$O = \{x \in U \mid x$ is a student who likes olives$\}$

Second, draw a Venn diagram with these sets and label each of the regions.

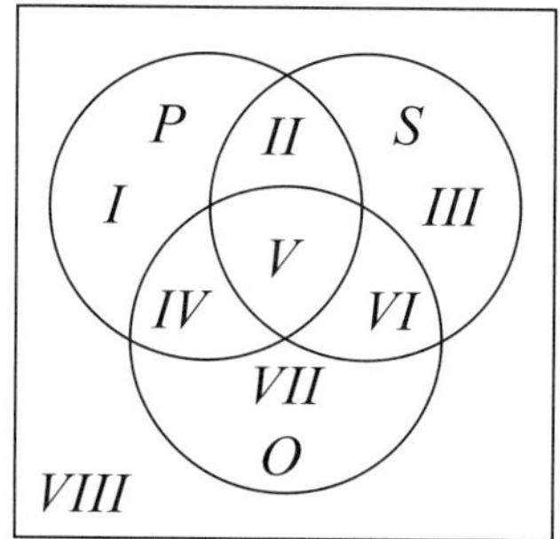

Convert each piece of information you are given into an equation relating the labeled regions. In doing so, you obtain the following equations:

- $I + II + IV + V = 190$
- $V + VI = 40$
- $IV + V + VI + VII = 130$
- $II + V = 150$
- $IV + V = 20$
- $II + III + V + VI = 200$
- $VIII = 5$
- $V = 5$

Substitute the values of regions you know, for example $V = 5$, into the remaining equations. Continue to do so with the new values you are able to find until you have found the values of each labeled region in the Venn diagram.

STEP	ORIGINAL EQUATION	SUBSTITUTED VALUES	FINAL VARIABLE VALUES
1	$VIII = 5$	no substitution needed	$VIII = 5$
2	$V = 5$	no substitution needed	$V = 5$
3	$V + VI = 40$	$5 + VI = 40$	$VI = 35$
4	$II + V = 150$	$II + 5 = 150$	$II = 145$
5	$IV + V = 20$	$IV + 5 = 20$	$IV = 15$
6	$I + II + IV + V = 190$	$I + 145 + 15 + 5 = 190$	$I = 25$
7	$IV + V + VI + VII = 130$	$15 + 5 + 35 + VII = 130$	$VII = 75$
8	$II + III + V + VI = 200$	$145 + III + 5 + 35 = 200$	$III = 15$

Notice that at the first pass you only know that $VIII = 5$ and $V = 5$. With this information you can find that $VI = 35$, $II = 145$, and $IV = 15$. Finally, you can find that $I = 25$, $VII = 75$, and $III = 15$. The final Venn diagram will look like the following:

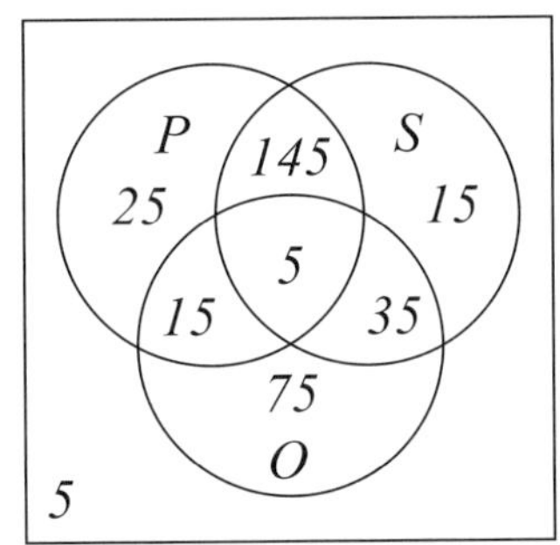

FURTHER EXPLORATIONS

For further explorations in the area of sets, please visit the URL

http://www.finitemathtutor.com/explore/chapter5/

EXERCISES

Exercise 5.1   Let $E = \{x \mid x$ is a letter in the word EASTER$\}$ and $H = \{x \mid x$ is a letter in the word HALLOWEEN$\}$. Find $n(E \cup H)$.

Exercise 5.2   Given the universal set $U = \{0, 1, 2, 3, 4, 5, 6, 7, 8, 9, 10\}$, $A = \{1, 2, 3, 4\}$, $B = \{2, 4, 6, 8, 10\}$, $C = \{0, 5, 10\}$, and $D = \{3, 7, 8, 10\}$, which of the following statements are true?

a. $\varnothing \in C$
b. $B \cap D = \{8, 10\}$
c. $C \subseteq A^c$
d. $3 \in D^c \cup A$
e. $\{2, 4\} \notin B$

Exercise 5.3   Let $A = \{$people who own trucks$\}$ and $B = \{$Texans$\}$. The intersection of Set A and Set B can best be described as which of the following?

a. Texans who do not own trucks
b. Texans or truck owners
c. Truck owners who do not live in Texas
d. Texans who own trucks

Shade the appropriate regions of a Venn diagram that represent the sets in Exercises 5.4–5.8.

Exercise 5.4 $A \cup C$

Exercise 5.5 $A \cap C$

Exercise 5.6 $A \cap B \cap C^c$

Exercise 5.7 $A^c \cup (B \cap C)$

Exercise 5.8 $A^c \cup B \cup C^c$

Exercise 5.9   A candle factory makes 80 different candles. Of the different candles, 40 are white and 34 are scented. If 70 candles are white or scented, how many candles are scented, but not white?

Exercise 5.10   A group of people is surveyed to find out which sports (football, baseball, or basketball) they like to watch.

- 50 people like watching football or baseball, but not basketball.
- 11 people like watching baseball and basketball.
- 13 people like watching basketball and football, but not baseball.
- 16 people don't like to watch any of these three sports.
- 47 people like to watch football.
- 34 people like to watch at least two of these sports.
- 3 people like to watch all three of these sports.
- 30 people like to watch basketball.

How many people were in the surveyed group?

SAMPLE QUIZ

Question 5.1   Given the sets $U = \{0, 1, 2, 3, 4, 5\}$, $A = \{0, 2, 4\}$, $B = \{1, 2, 3, 4, 5\}$ and $C = \{2, 3, 4\}$ which of the following statements are true?

a. $\varnothing \in U$
b. $A \cup B = U$
c. $C \subset B$
d. $C^c \cap A = \varnothing$

Question 5.2   The universal set $U = \{0, 1, 2, 3, 4, 5, 6, 7, 8, 9\}$ and $A = \{$even integers between 1 and 7$\}$. Find $n(A)$ and $n(A^c)$.

Question 5.3   Given that $n(A) = 12$, $n(B) = 10$, $n(A \cup B) = 18$, and $n(A^c \cap B^c) = 5$, find $n(A \cap B)$, $n(A^c \cap B)$, and $n(A \cup B^c)$.

Question 5.4   Given $n(A) = 40$, $n(B) = 35$, $n(A \cup B) = 55$, and $n(U) = 100$, find $n(A \cap B)$, $n(A^c \cap B)$, and $n(A \cup B^c)$.

Question 5.5   Sixty grocery shoppers were surveyed. This week, 45 shoppers bought eggs, 8 bought steak, and 10 bought neither steak nor eggs. Find the number of shoppers who bought

a. steak or eggs.
b. only steak.
c. steak and eggs.

For Questions 5.6 and 5.7, shade the appropriate regions of a Venn diagram representing the given sets.

Question 5.6 $A \cap B^c \cap C$

Question 5.7 $A^c \cap (B^c \cup C)$

Question 5.8   Fill in a Venn diagram corresponding to the following information:

$n(U) = 50$ $\qquad\qquad$ $n(A \cap B^c \cap C) = 1$
$n(A \cap B \cap C) = 2$ $\qquad\qquad$ $n((B \cup C)^c) = 10$
$n(A \cap B) = 5$ $\qquad\qquad$ $n((A \cap B^c \cap C) \cup (A^c \cap B \cap C)) = 11$
$n(A) = 12$ $\qquad\qquad$ $n(A^c \cap B \cap C^c) = 5$

Question 5.9   Given that

$n(B) = 20$ $\qquad\qquad$ $n(A^c \cap B^c \cap C) = 10$
$n(A \cap B \cap C) = 4$ $\qquad\qquad$ $n(A \cup B \cup C) = 37$
$n(A \cap B) = 7$ $\qquad\qquad$ $n((A \cap B \cap C^c) \cup (A^c \cap B \cap C)) = 14$
$n(A \cap B^c \cap C^c) = 8$ $\qquad\qquad$ $n((A \cup B)^c) = 17$

find the following:

a. $n(A \cup B)$ $\qquad\qquad$ **b.** $n(C)$ $\qquad\qquad$ **c.** $n(U)$

Question 5.10　A group of children was asked what they like on their hamburgers from a choice of lettuce, pickles, and tomatoes. The following results were found:

- 8 children liked only pickles
- 4 children liked all three items
- 23 children liked exactly two items
- 7 children liked only pickles and tomatoes
- 10 children liked pickles and lettuce
- 30 children liked tomatoes
- 17 children liked lettuce and not tomatoes
- 20 children did not like lettuce or tomatoes

a. How many children were surveyed?
b. How many children liked pickles?
c. How many children liked at least two of the items?

Counting and Probability

6

As we saw in Chapter 5, one very important set operation is counting the number of elements in a set. This procedure can also be applied to subsets, complements, unions, and intersections of sets. In this chapter we will develop basic counting methods and explore their relationships to probability.

6.1 MULTIPLICATION PRINCIPLE

Often, we must perform a succession of activities and look at the possible results of our actions. To be specific, suppose there are two tasks to accomplish, for example, rolling a six-sided die and tossing a two-sided coin. There are 6 possible ways for the die to land, and 2 ways for the coin to land. How many different possibilities are there for both tasks to occur? In this particular example, we can list all the possibilities using a **tree diagram** (Figure 6.1).

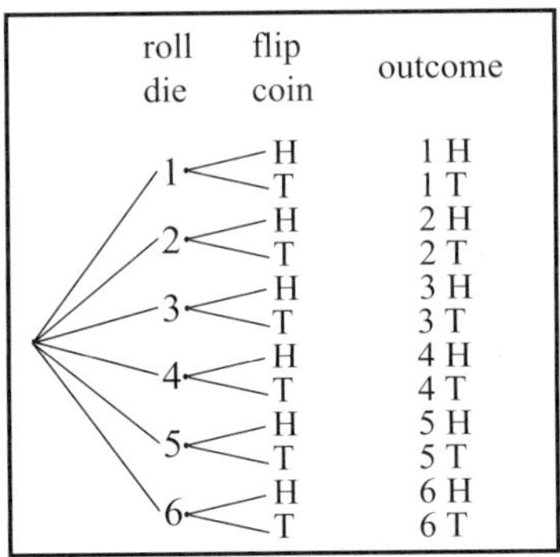

Figure 6.1   Tree Diagram for a One Die and One Coin Experiment

There are $6 \cdot 2 = 12$ possibilities in all, six where the coin lands heads, and six where the coin lands tails. The number of possibilities is the *product* of the ways in which each task can be accomplished. This is a very general result, and is called the **multiplication principle**.

> If there are k tasks, and N_1 ways to accomplish task 1, N_2 ways to accomplish task 2, and so on, then there are
>
> $$N = N_1 \times N_2 \times \cdots \times N_k$$
>
> ways to accomplish all k tasks.

In certain counting problems you will need to provide a list of all possible **outcomes** for an **experiment**. This list is called the **sample space**. For this coin and die experiment, the sample space is

$$S = \{1H, 1T, 2H, 2T, 3H, 3T, 4H, 4T, 5H, 5T, 6H, 6T\}$$

A tree diagram is a useful way to find the outcomes in the sample space. However, in many cases a list of the outcomes is not needed; only the number of outcomes in the sample space is required. In this case it is useful to draw a blank for each task to be completed and label (if possible) each empty blank. Then determine the number of ways that each task can be completed and perform the multiplication:

$$\underset{\text{Roll Die}}{\underline{}} \times \underset{\text{Toss Coin}}{\underline{}} \quad \rightarrow \quad \underset{\text{Roll Die}}{\underline{6}} \times \underset{\text{Toss Coin}}{\underline{2}} \quad = \quad 12$$

Example 6.1   A coffee shop offers lattes in four different sizes with eight different kinds of flavored syrup and a choice of skim or whole milk. How many different lattes are possible?

Solution

We have three tasks to complete: choose a size, choose a syrup, and choose a type of milk. We can use the multiplication principle to determine the number of possible lattes:

$$\underline{} \times \underline{} \times \underline{} \quad \rightarrow \quad \underline{\;4\;} \times \underline{\;8\;} \times \underline{\;2\;} = 64$$
$$\text{Size} \quad \text{Syrup} \quad \text{Milk} \qquad \text{Size} \quad \text{Syrup} \quad \text{Milk}$$

There are 64 different lattes possible. 

Example 6.2 An experiment consists of flipping a fair coin twice. How many outcomes are in the sample space?

Solution

When finding a sample space, we want to be sure that each of the outcomes in our list has an equal chance of occurring (a **uniform sample space**). Each coin has two possible outcomes,

$$\underline{} \times \underline{} \quad \rightarrow \quad \underline{\;2\;} \times \underline{\;2\;} = 4$$
$$\text{Toss Coin 1} \quad \text{Toss Coin 2} \qquad \text{Toss Coin 1} \quad \text{Toss Coin 2}$$

Therefore the sample space has 4 outcomes. 

 **Note:** *You should now complete Activity A in the Counting and Probability Module.*

It is important to note that the *order* in which we consider the tasks in these first two examples is unimportant. Considering whether the order of the tasks, and their results, is important brings us to the next topic.

6.2 PERMUTATIONS

Suppose we have N objects and we take r of them at a time, where the *order matters*. How many possible **permutations** are there? For example, suppose we have a set of 10 letters $\{a, b, c, d, e, f, g, h, i, j\}$ and we choose 4 different letters from this set to make a 4-letter word. How many different possible words can be created?

If the letters can be *repeated*, we proceed as follows: There are 10 possible choices for the first letter, 10 possible choices for the second, 10 for the third, and 10 for the fourth. By the multiplication principle, this gives $10{,}000 = 10 \times 10 \times 10 \times 10$ different words.

If the letters *cannot* be repeated, we proceed as follows: There are 10 possible choices for the first letter, but only 9 for the second – since one letter has been chosen. There are 8 possible letters for the third, and 7 for the fourth. By the multiplication principle, there are $5{,}040 = 10 \times 9 \times 8 \times 7$. This particular type of product comes up so often, we have developed the shorthand **factorial** notation as part of it. $N!$ (read "N-factorial") is defined as

$$N! = N \times (N - 1) \times (N - 2) \times \cdots \times (2) \times (1)$$

It is the product of the first N integers. $0!$ is equal to 1. We can write the result of the previous example as

$$
\begin{aligned}
5040 &= 10 \times 9 \times 8 \times 7 \\
&= \frac{10 \times 9 \times 8 \times 7 \times 6 \times 5 \times 4 \times 3 \times 2 \times 1}{6 \times 5 \times 4 \times 3 \times 2 \times 1} \\
&= \frac{10!}{6!}
\end{aligned}
$$

Any product of sequential integers can be written as a ratio of factorials.

The number of ways in which N objects can be chosen, r at a time, where order *does* matter is equal to

$$P(N,r) = \frac{N!}{(N - r)!}$$

 **Example 6.3** You have nine different potted plants to arrange on a shelf. How many different arrangements of all nine plants are possible? How many different arrangements of three of the nine plants?

Solution

This is a permutation since we are choosing different items from a finite set and arranging them. For all nine plants,

$$P(9,9) = \frac{9!}{(9-9)!} = 362,880$$

For three of the nine plants,

$$P(9,3) = \frac{9!}{(9-3)!} = 504$$

So there are 362,880 ways to arrange all nine plants and 504 ways to arrange three of the nine plants.   ❖

In these examples, the order of the items is important. For example, when arranging letters, the word "bead" is different than the word "deab" (especially since this last arrangement is not even an English word). If we ask what happens if the order is not important, we come to our next topic.

6.3   COMBINATIONS

Suppose we have N objects, and we take r of them at a time, but the order is not important. How many possible **combinations** are there? For example, given a standard deck of 52 cards (four suits, 13 ranks), and we choose five cards, how many different five-card hands are there? In the case of cards, it does not matter the order in which we hold them in our hand, just the set of cards itself. This is analogous to the difference between a **set** and a **list**. In a set, the order is not important – the set $\{1, 2\}$ is the same as the set $\{2, 1\}$. In a list, however, the order is important. The ordered pair $(1, 2)$ is different from $(2, 1)$. Returning to our example, how many different 5-card hands are there?

We have a choice of all 52 cards for the first card, 51 cards for the second, 50 for the third, 49 for the fourth, and 48 for the fifth. If we had only two cards in our hand, there are $2 = 2 \cdot 1$ equivalent ways to arrange the cards. If we have three cards in our hand, there are $6 = 3 \cdot 2 \cdot 1$ ways:

$$\{1^{st}, 2^{nd}, 3^{rd}\}, \{1^{st}, 3^{rd}, 2^{nd}\}, \{2^{nd}, 1^{st}, 3^{rd}\}, \{2^{nd}, 3^{rd}, 1^{st}\}, \{3^{rd}, 1^{st}, 2^{nd}\}, \{3^{rd}, 2^{nd}, 1^{st}\}$$

If we have four, there are $24 = 4 \cdot 3 \cdot 2 \cdot 1$ ways, and for five cards $120 = 5 \cdot 4 \cdot 3 \cdot 2 \cdot 1$. In the case of five cards, we must divide the total number of permutations by the number of equivalent combinations

$$\frac{52 \times 51 \times 50 \times 49 \times 48}{5 \times 4 \times 3 \times 2 \times 1} = \frac{52!}{47! \times 5!}$$

The number of ways one can take N objects, r at a time, where order is *not* important, is equal to

$$C(N,r) = \frac{N!}{(N-r)!r!} = \frac{P(N,r)}{r!}$$

Example 6.4   You have a bag of 10 apples and choose two apples for a snack. How many different two-apple snacks are possible?

Solution

The order that the apples are chosen doesn't matter, so this will be a combination of 10 objects taken two at a time,

$$C(10,2) = \frac{10!}{(10-2)!2!} = 45$$

So there are 45 different ways to choose two apples for a snack.   ❖

6.4   MIXED PROBLEMS

The multiplication principle is deceptively simple to state and can be quite difficult to apply in situations where there are restrictions, groups, or identical items. The following examples illustrate some of the more typical applications.

Example 6.5   You have a bag of 4 blue, 3 red, 6 green, and 1 black marbles. How many distinguishable ways can these 14 marbles be arranged?

Solution

When arranging items, some of which are identical, care must be taken to count only arrangements that *look different*. If two identical red marbles are swapped, the arrangement will still look the same. However, we start with the number of ways to arrange the items as if they were all different. In this case there will be 14! ways to arrange the marbles. There are 3 red marbles and any time these are switched among each other it will not change the appearance of the arrangement; there are 3! ways to do this. There are 4! ways to arrange the blue marbles and 6! ways to arrange the green marbles.   So the number of distinguishable arrangements will be

$$N = \frac{14!}{4! \times 3! \times 6!} = 840,840$$

❖

Example 6.6   A class of 12 children (6 boys and 6 girls) goes to lunch. How many ways can these children be seated in a row if boys and girls must alternate? How many ways if boys must sit together and girls must sit together?

Solution

Using the multiplication principle, we have 12 tasks to perform. When seating alternately, there are 12 choices for who sits down first. If a boy is seated, then one of the 6 girls must be chosen to sit next to him, then one of the 5 remaining boys, then one of the 5 remaining girls, and so on. In all,

$$N = 12 \cdot 6 \cdot 5 \cdot 4 \cdot 4 \cdot 3 \cdot 3 \cdot 2 \cdot 2 \cdot 1 \cdot 1 = 1,036,800$$

To seat the boys together and the girls together is similar. There are 12 choices for the first child to sit down. If a boy is chosen first, the next child is chosen from the 5 remaining boys, and so on until all the boys are seated. Then the 6 girls are seated.

$$N = 12 \cdot 5 \cdot 4 \cdot 3 \cdot 2 \cdot 1 \cdot 6 \cdot 5 \cdot 4 \cdot 3 \cdot 2 \cdot 1 = 1,036,800$$

We find in both cases that there are more than one million ways to seat these children!   ❖

Example 6.7   You have 2 black and 5 blue jellybeans. How many ways can these 7 jellybeans be arranged if two black jellybeans cannot be next to each other?

Solution

We can arrange the 5 blue jellybeans so that there is a space between each:

$$\_\,B\,\_\,B\,\_\,B\,\_\,B\,\_\,B\,\_$$

There are in all 6 spaces to place a black jellybean. Since order doesn't matter, there are $C(6, 2) = 15$ ways to place the black jellybeans so that they are not together.   ❖

Example 6.8   A group of 12 workers has 5 men and 7 women. A delegation of 3 will be chosen to represent the group. How many ways can a delegation of 3 be chosen with at least two women?

Solution

Start with the "at least two women" statement. That is going to mean delegations of exactly two women or delegations of exactly three women. These are disjoint sets (a delegation cannot have exactly two and exactly three women at the same time).

To have exactly two women, we have to choose any 2 of the 7 women, which can happen $C(7, 2) = 21$ ways. One person remains to be chosen from the five men, which can be done $C(5, 1) = 5$ ways. In all, there will be $21 \cdot 5 = 65$ ways to have a delegation with exactly two women.

To have exactly three women, we have to choose any 3 of the 7 women, which can happen $C(7, 3) = 35$ ways. No one remains to be chosen from the five men, which can be done $C(5, 0) = 1$ way. In all, there will be $35 \cdot 1 = 35$ ways to have a delegation with exactly three women. In all, the number of ways to have at least two women on the delegation is $65 + 35 = 100$ ways.   ❖

Note: *You should now complete Activity B in the Counting and Probability Module.*

6.5   ELEMENTARY PROBABILITY

We can use these elementary counting techniques to help us find the probability of certain types of **events** occurring. Consider the example of throwing a single die. What is the probability of the die landing with the 2 showing? If the die is "fair" — that is, the probability of each number occurring is equally likely — then

$$P(1) = P(2) = P(3) = P(4) = P(5) = P(6) = \frac{1}{6}$$

If we actually were to roll a die six thousand times, we would expect to see that each number would occur *approximately* one thousand times. It would be unrealistic to expect that each number occurs *exactly* one thousand times, nor would we expect that a number would come up two thousand times (twice as often). The percentage of times that a number occurs is called the **relative frequency** or **empirical probability**. It is based on what *actually* occurs, rather than a *theoretical* analysis of what should occur. As the number of trials increases, one would expect that the theoretical and empirical probabilities would agree! The language of probability includes the following terms:

- **Experiment:** An activity with an observable outcome.
- **Outcome:** Something that happens when an activity is done.
- **Sample space:** The set of all possible outcomes of an experiment.
- **Event:** A subset of the sample space.

All theoretical probability calculations are based on counting and the laws of probability (Table 6.1)

- $P(E) \geq 0$ for any event E in the sample space.
- $P(\emptyset) = 0$ and $P(S) = 1$ where $\emptyset$ is the empty set and S is the entire sample space
- $P(E \cap F) = P(E) + P(F)$ if E and F are disjoint, that is if $E \cap F = \emptyset$.
- $P(E \cap F) = P(E) + P(F) + P(E \cap F)$ for any events E and F in the sample space (the Union Rule)

Table 6.1   Laws of Probability

Example 6.9   A cup contains one red, one yellow, and one blue marble. A single marble is drawn at random from the cup. What is the sample space for this experiment? How many different events are possible for this experiment?

Solution

The sample space has three outcomes, $S = \{blue, red, yellow\} = \{b, r, y\}$. The subsets of this set (see Chapter 5 for details about subsets) are

$$\varnothing, \{b\}, \{r\}, \{y\}, \{b, r\}, \{b, y\}, \{r, y\}, \text{ and } \{b, r, y\}$$

So there are 8 possible events in this experiment.   ❖

Example 6.10   Two fair six-sided dice are rolled.

a. What is the probability that the sum of the two dice is 8?
b. What is the probability that the sum is 8 and the roll was a double (both die showing the same number)?
c. What is the probability that the sum is 8 or the roll was a double (both die showing the same number)?

Solution

Start with the sample space for rolling two fair six-sided dice. There are 36 outcomes in the sample space since each die has 6 possible outcomes:

1~1	2~1	3~1	4~1	5~1	6~1
1~2	2~2	3~2	4~2	5~2	6~2
1~3	2~3	3~3	4~3	5~3	6~3
1~4	2~4	3~4	4~4	5~4	6~4
1~5	2~5	3~5	4~5	5~5	6~5
1~6	2~6	3~6	4~6	5~6	6~6

a. There are 5 outcomes that have a sum of 8 (2~6, 3~5, 4~4, 5~3, and 6~2). Therefore the probability is $\frac{5}{36}$.

b. There is one roll that is both a double and a sum of 8 (4~4) and therefore the probability is $\frac{1}{36}$.

c. The "or" used in probability is the inclusive or, and so we need to count the outcomes that are doubles, sums of 8, and both. There are a total of 10 outcomes (1~1, 2~2, 3~3, 4~4, 5~5, 6~6, 2~6, 3~5, 5~3, 6~2; notice the 4~4 outcome was not counted twice!) and the probability is $\frac{10}{36}$ or $\frac{5}{18}$. ❖

The Counting and Probability Module contains an applet to simulate drawing balls from an urn. We can start with 10 balls in the urn: 3 red (light gray), 2 green (medium gray), and 5 blue (dark gray) balls. The legs of the tree diagram show the theoretical probability that a ball is chosen and to the right of the ball is shown the empirical probability that a ball is chosen. Figure 6.2 shows the urn before any balls are chosen.

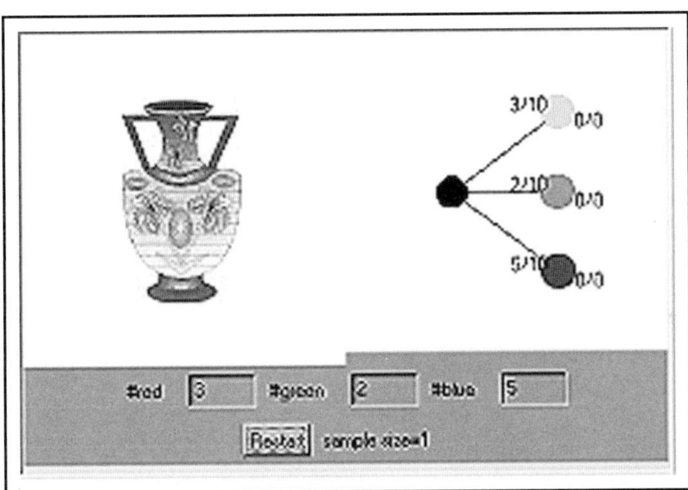

Figure 6.2   Urn Applet

When the urn is clicked, a ball is chosen at random. Figure 6.3 shows what might happen after 10 balls are drawn (a ball is replaced after it is drawn so there are always 10 balls to choose from). Figure 6.4 shows what might happen after 100 balls are drawn. We see that the empirical and theoretical probabilities become closer after the experiment is repeated more times.

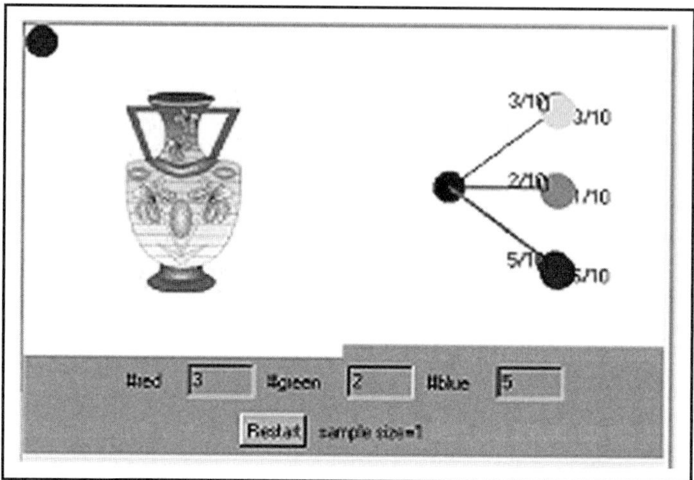

Figure 6.3   Urn After 10 Balls Are Drawn

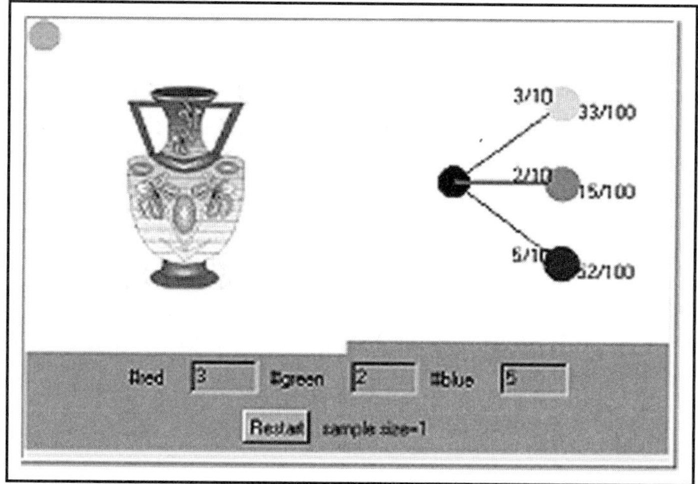

Figure 6.4   Urn After 100 Balls Are Drawn

 **Note:** *You should now complete Activity C in the Counting and Probability Module.*

FURTHER EXPLORATIONS

For further explorations in the area of counting and probability, please visit the URL

http://www.finitemathtutor.com/explore/chapter6/

EXERCISES

Exercise 6.1   An ice cream shop offers a mini-sundae with a choice of 27 different ice cream flavors, 6 different sauces, and 10 different toppings. How many different mini-sundaes are possible?

Exercise 6.2   There are 8 different stuffed animals to arrange on a shelf. How many different arrangements of these stuffed animals are possible?

Exercise 6.3   A group of 9 children and 9 adults goes to a movie. How many different ways can these 18 people be seated if children and adults must alternate?

Exercise 6.4   You have 5 identical silver beads, 3 identical gold beads, and 6 identical copper beads. How many distinguishable ways can these 14 beads be strung on a string?

Exercise 6.5   The letters a, b, c, d, e, f, and g are written on 7 identical pieces of paper and placed in a bag. A letter is drawn at random from the bag and a single fair coin is tossed. How many different outcomes are in the uniform sample space of this experiment?

Exercise 6.6   You have 6 blue marbles and 4 green marbles. How many arrangements of these 10 marbles are possible if 2 green marbles cannot be next to each other?

Exercise 6.7   A cup contains 4 wooden tiles with the letters w, x, y, and z on them. A single tile is drawn at random and the letter written on the tile is noted. How many events are possible in this experiment?

Exercise 6.8   You purchase a 24-pack of soda. You are told that 3 of the cans contain a winning prize. You choose a sample of 3 cans at random. What is the probability that you have at least one winning prize in your sample?

Exercise 6.9   A math class has 19 students. Nine of the students are freshmen, 6 are sophomores, 3 are juniors, and 1 is a senior. A group of 4 students is chosen at random from the class. What is the probability that the group has exactly one freshman, one sophomore, one junior, and one senior?

Exercise 6.10   You are dealt a single card from a standard well-shuffled deck of 52 cards. What is the probability it is a heart or a queen?

SAMPLE QUIZ

Question 6.1   A cafeteria special plate lets you choose one of six entrees, one of eight salads, and one of three side dishes. How many different special plates are possible?

Question 6.2   You have seven different perfume bottles to arrange on a shelf. How many different arrangements of these bottles are possible?

Question 6.3   You have five different green books and five different yellow books. How many ways can these books be arranged on a shelf if the colors must alternate?

Question 6.4   How many different "words" can be made from the letters in the word *workbook*?

Question 6.5   You roll a six-sided die and a twelve-sided die. How many outcomes are in the uniform sample space for this experiment?

Question 6.6   You have 3 A's and 8 B's. How many arrangements are possible if 2 A's cannot be next to each other?

Question 6.7   A bowl contains a red ball, a blue ball, and a yellow ball. A single ball is drawn at random from the bowl. How many events are possible in this experiment?

Question 6.8   You have a carton of 12 eggs. Three of the eggs have double yolks. A sample of two is chosen at random for breakfast. What is the probability that your sample has at least one double yolk egg?

Question 6.9   A cup has 4 red, 5 white, and 3 blue marbles. A sample of 3 is chosen at random. What is the probability that you have exactly one of each color?

Question 6.10   A pair of fair six-sided dice is rolled. What is the probability that a sum of 8 or at least one 5 is shown uppermost?

Conditional Probability 7

We introduced the basic concepts of probability in Chapter 6. The main lesson was that for uniform probability distributions, the theoretical probability of an event E occurring is given by the ratio

$$P(E) = \frac{n(E)}{n(S)}$$

where $n(E)$ is the number of ways that the event E can occur, and $n(S)$ is the number of elements in the sample space, S. This reduces the calculation of probabilities to simple counting and elementary arithmetic. The process becomes more challenging when events are linked. For example, consider the following question:

An urn has 5 white balls and 6 black balls in it. What is the probability that the *second* ball chosen from the urn is black?

Assuming that the first ball is not placed back into the urn before the second ball is drawn, it is clear that the probability of choosing a black ball is higher if the first ball is white, since there will then be a higher ratio of black balls to white balls in that case. We expect the probability of choosing a black ball to decrease if the first ball drawn is black, since the ratio of white balls to black balls then increases. This illustrates a certain subtlety in these types of problems: If we *know* the color of the first ball chosen, that affects the probabilities for the second ball. On the other hand, if we draw two balls in repeated trials, and just record the color of the second one, the empirical probability should approach some definite limit. We discuss the probabilities of related events in this chapter.

7.1   CONDITIONAL PROBABILITY

Example 7.1   Consider the previous question of drawing two balls from an urn with 5 white balls and 6 black balls, where the first ball is not replaced.

a. What is the probability that the second ball drawn is black, given that the first ball drawn was white?

b. What is the probability that the second ball drawn is black, given that the first ball drawn was black?

Solution

a. Since the first ball drawn is white, there are 4 white balls and 6 black balls left in the urn. Let B be the event of drawing a black ball on the second draw. The probability of drawing a black ball second, $P(B)$, in this scenario is given by

$$P(B) = \frac{n(B)}{n(S)} = \frac{6}{10} = 0.6$$

b. Since there are now 5 white balls and 5 black balls left, the probability of drawing a black ball in this scenario is given by

$$P(B) = \frac{n(B)}{n(S)} = \frac{5}{10} = 0.5$$

Notice that $P(B)$ changed according to what happened on the first draw.   

Given two events E and F in a sample space, we can ask what the probability is that F occurs *given that E occurs*. We introduce special notation for this:

$$P(F \mid E) = \text{probability that } F \text{ occurs given that } E \text{ occurs}$$

We can find a formula that relates to the probabilities $P(F \mid E)$ and $P(E)$. Consider the Venn diagram shown in Figure 7.1.

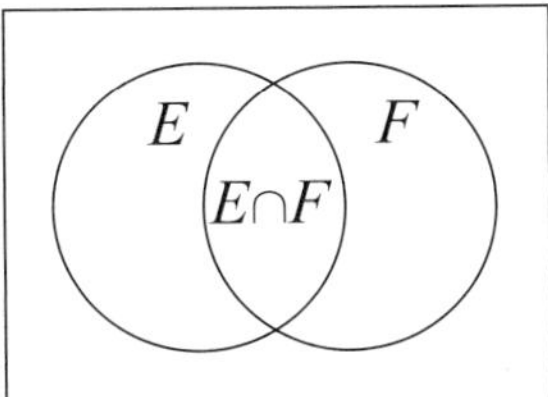

Figure 7.1   Overlapping Events

Since the event E has already occurred, we only need to consider the events where both E and F occur. Therefore, we have the formula

$$P(F\,|\,E) = \frac{n(E \cap F)}{n(E)}$$

This says that the **conditional probability** is the ratio of the number of events where *both* E and F occur, to the number of events where E has occurred. It is also useful to rewrite this identity as

$$P(F\,|\,E) = \frac{n(E \cap F)}{n(E)} = \frac{\dfrac{n(E \cap F)}{n(S)}}{\dfrac{n(E)}{n(S)}} = \frac{P(E \cap F)}{P(E)}$$

which gives the conditional probability as a ratio of two probabilities. An equivalent expression is

$$P(E \cap F) = P(E) \cdot P(E\,|\,F)$$

Example 7.2   Two balls are drawn one at a time from an urn with 5 white balls and 6 black balls, where the first ball is not replaced.

a. What is the probability that the second ball drawn is white, given that the first ball drawn was white?
b. What is the probability that the second ball drawn is white, given that the first ball drawn was black?
c. What is the probability that the second ball drawn is white?

Solution

This problem closely resembles the problem in Example 7.1; however, we are going to solve this problem using a tree diagram and conditional probabilities.

For our problem, let W be the event that the first ball drawn is white, W^c be the event that the first ball drawn is not white (that it is black), w be the event that the second ball drawn is white, and w^c be the event that the second ball drawn is not white (that it is black).

Recall that tree diagrams can be used to illustrate the possible outcomes from an experiment by showing each stage of the experiment separately and how the stages relate to one another. We can further illustrate how the stages are related by labeling each branch with its associated probability.

The first stage of our experiment is to draw a ball from the urn. There are two possible outcomes — we draw a white ball or we don't. Since this is the first stage, no events have occurred previously, and therefore the probabilities associated with the events at this stage are pure probabilities (Figure 7.2).

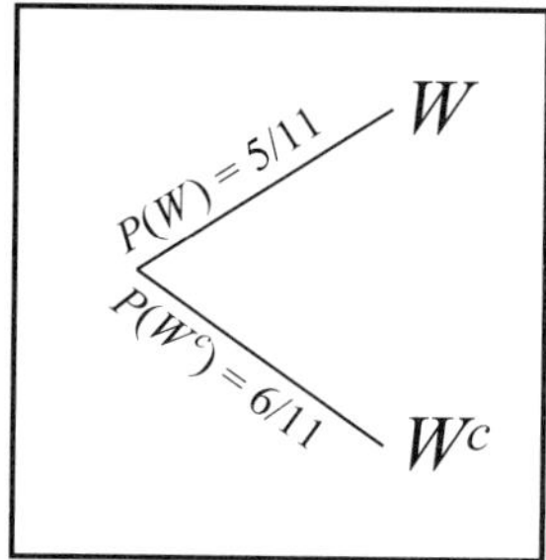

Figure 7.2   Tree Diagram of the First Stage

The second stage in our experiment is to draw a second ball from the urn. Again, there are two possible outcomes – we draw a white ball or we don't. However, at this stage, we have already had an event occur. In other words, we are *given* information before we begin this stage. Therefore, in the second stage of this experiment, and all subsequent stages if they were to exist, the probabilities associated with the events are conditional probabilities (Figure 7.3).

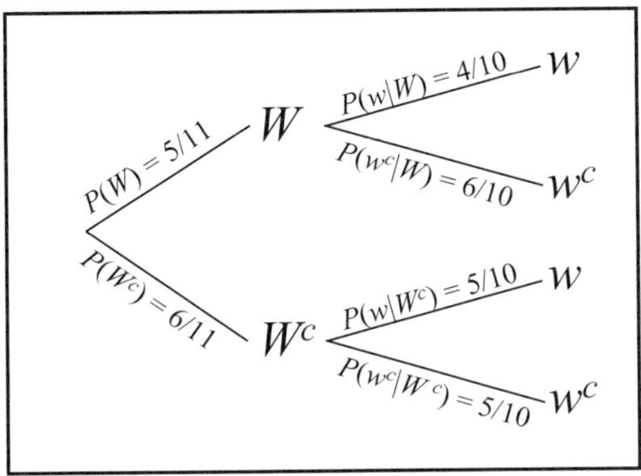

Figure 7.3   Tree Diagram of the Entire Experiment

It is important to notice that the probabilities of all of the branches stemming from the same origin add to 1.

$$P(W) + P(W^c) = 1; \qquad P(w \mid W) + P(w^c \mid W) = 1; \qquad P(w \mid W^c) + P(w^c \mid W^c) = 1$$

This follows from the fact that $P(S) =$ the probability of all possible outcomes $= 1$, and at each origin point, all of the possibilities from that point should be represented by a branch stemming from the point.

By the same argument, the probability of all of the possible outcomes from the entire experiment should add to 1, also. All of the possible outcomes from the entire experiment are represented by all of the paths through the tree diagram that start at the leftmost point on the tree and end at the rightmost points. For our problem, we have the following disjoint (mutually exclusive) outcomes: $\{Ww, Ww^c, W^cw, W^cw^c\}$. Therefore,

$$\begin{aligned}
1 &= P(Ww) + P(Ww^c) + P(W^cw) + P(W^cw^c) \\
&= P(W \cap w) + P(W \cap w^c) + P(W^c \cap w) + P(W^c \cap w^c) \\
&= P(W) \cdot P(w|W) + P(W) \cdot P(w^c|W) + P(W^c) \cdot P(w|W^c) + P(W^c) \cdot P(w^c|W^c)
\end{aligned}$$

Notice that to find the probabilities of the intersections of events by using the given identity from this chapter just amounts to multiplying the probabilities of the branches you use for a particular outcome.

Using the tree diagram in Figure 7.3, we can now answer the questions in Example 7.2.

a. We are looking for $P(w \mid W)$. Notice that this probability is a value on our tree, which is equal to 0.40.

b. We are looking for $P(w \mid W^c)$. Again, notice that this is a value on our tree, which is equal to 0.50.

c. We are looking for $P(w)$. Since we are looking for the pure probability of an event that is not in the first stage, we have to consider all of the possible ways that the event can occur. Looking at the tree, it is easy to see that there are two disjoint ways for the second ball to be white – draw two whites or draw a black and then a white. Therefore, we have

$$P(w) = P(W \cap w) + P(W^c \cap w)$$
$$= P(W) \cdot P(w \mid W) + P(W^c) \cdot P(w \mid W^c)$$
$$= \left(\frac{5}{11}\right)\left(\frac{4}{10}\right) + \left(\frac{6}{11}\right)\left(\frac{5}{10}\right)$$
$$= \frac{5}{11} \approx .4545$$

That is, the probability of choosing a white ball second is $\frac{5}{11}$.   ❖

Example 7.3   Two fair six-sided dice are rolled. What is the probability that a sum of 10 is rolled, given that the first die rolled shows a number greater than 4?

Solution

Let F be the event that a sum of 10 is rolled and let E be the event that the first die rolled shows a number greater than 4. Therefore, you are looking for

$$P(F|E) = \frac{P(F \cap E)}{P(E)}$$

To find the probabilities we need, it is useful to see all of the possible outcomes from this experiment. In this experiment, it is easier to see the outcomes in a table, rather than in a tree diagram (Figure 7.4).

$$
\begin{array}{cccccc}
1\sim1 & 2\sim1 & 3\sim1 & 4\sim1 & 5\sim1 & 6\sim1 \\
1\sim2 & 2\sim2 & 3\sim2 & 4\sim2 & 5\sim2 & 6\sim2 \\
1\sim3 & 2\sim3 & 3\sim3 & 4\sim3 & 5\sim3 & 6\sim3 \\
1\sim4 & 2\sim4 & 3\sim4 & 4\sim4 & 5\sim4 & 6\sim4 \\
1\sim5 & 2\sim5 & 3\sim5 & 4\sim5 & 5\sim5 & 6\sim5 \\
1\sim6 & 2\sim6 & 3\sim6 & 4\sim6 & 5\sim6 & 6\sim6 \\
\end{array}
$$

Figure 7.4   Outcomes when Rolling Two Six-Sided Dice

Since each of the 36 outcomes is equally likely, we can see that the intersection, where the outcome is both a sum of 10 and the first die has a number greater then 4 (shown in dark highlighting in Figure 7.4) has a probability of

$$
P(E \cap F) = P(5 \sim 5, 6 \sim 4) = \frac{2}{36}
$$

and the given event (shown highlighted in the last two columns of Figure 7.4) has a probability of

$$
P(E) = P(5\sim1, 5\sim2, 5\sim3, 5\sim4, 5\sim5, 5\sim6, 6\sim1, 6\sim2, 6\sim3, 6\sim4, 6\sim5, 6\sim6) = \frac{12}{36}
$$

so

$$
P(F \mid E) = \frac{P(F \cap E)}{P(E)} = \frac{\left(\dfrac{2}{36}\right)}{\left(\dfrac{12}{36}\right)} = \frac{2}{12} = \frac{1}{6}
$$

Notice that $P(F \mid E)$ is different than $P(F) = P(4\sim6, 5\sim5, 6\sim4) = \frac{3}{36}$.   ❖

 **Note:** *You should now complete Activity A in the Conditional Probability Module.*

7.2   INDEPENDENT EVENTS

When we say two events are **independent**, intuitively we mean that the probability of one event does not affect the other. In other words, the probability of event E occurring is the same whether or not the event F occurs. This means that

$$P(E) = P(E|F)$$

or

$$P(E) = P(E \mid F) = \frac{P(E \cap F)}{P(F)}$$

or, finally:

$$\boxed{P(E \cap F) = P(E) \cdot P(F)}$$

Two events are **independent events** if and only if the probability of E *and* F occurring is the product of the probabilities.

This can be expanded to account for more than just two events by the following:

$$\boxed{\begin{array}{c} \text{If } E_1, E_2, \ldots, E_N \text{ are independent events, then} \\ P(E_1 \cap E_2 \cap \ldots \cap E_N) = P(E) \cdot P(E_2) \ldots \cdot P(E_N) \end{array}}$$

 **Example 7.4**   Suppose $P(A) = 0.6$, $P(B) = 0.4$ and $P(A \cap B) = 0.24$. Are A and B independent events?

Solution

To determine whether or not A and B are independent, we need to check that

$$P(A \cap B) = P(A) \cdot P(B)$$

Notice that we are given two of these three values — all we need to find is $P(A)$. Recall that $P(A^c) = 1 - P(A)$. Therefore, $P(A) = 1 - 0.6 = 0.4$. Checking the independence identity gives us $0.24 \neq (0.4)(0.4) = 0.16$, and so A and B are not independent events.   ❖

Example 7.5   An undercover reporter wants to find out the accuracy of doctors' diagnoses in the area. After having hundreds of patients with known illnesses see three doctors (Dr. Smith, Dr. Long, and Dr. Hart), the reporter finds that Dr. Smith gives a correct diagnosis 80% of the time, Dr. Long is correct only 65% of the time, and Dr. Hart is correct 99% of the time. Assuming the three doctors diagnose patients independently from one another, what is the probability that a person will be given a correct diagnosis by all three of these doctors?

Solution
Let S be the event that Dr. Smith gives a correct diagnosis, L be the event that Dr. Long gives a correct diagnosis, and H be the event that Dr. Hart gives a correct diagnosis. The information from the reporter tells us that

$$P(S) = 0.80,\ P(L) = 0.65,\ \text{and}\ P(H) = 0.99$$

Finding the probability that all three doctors diagnose the person correctly is equivalent to finding $P(S \cap L \cap H)$. Since the doctors act independently, that gives us

$$P(S \cap L \cap H) = P(S) \cdot P(L) \cdot P(H)$$
$$= (0.80)(0.65)(0.99)$$
$$= 0.5148$$

This means that there's just above a 50% chance that all of these doctors will give a correct diagnosis — maybe we should look for some different doctors!   ❖

Note: *You should now complete Activity B in the Conditional Probability Module.*

7.3   BAYES' THEOREM

Consider a two-stage experiment with only one possible outcome at each stage. A tree diagram representing such an experiment is given in Figure 7.5.

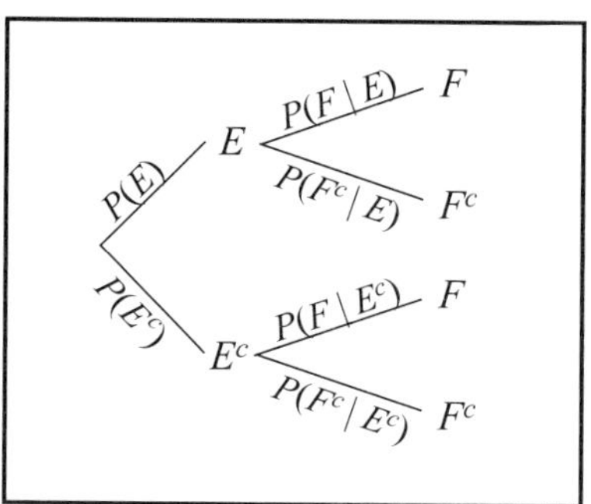

Figure 7.5   Binary Tree Diagram

We have seen how knowing the outcome at the *first* stage affects the probabilities of the outcomes at the second stage. What happens if we already know the outcomes of the *second* stage? For instance, using our two-stage experiment, what is the probability that E occurred at the first stage, given that F occurs on the second? This amounts to finding the conditional probability $P(E \mid F)$, which is not a value labeled on our tree. By our conditional probability identities, we have

$$
\begin{aligned}
P(E \mid F) &= \frac{P(E \cap F)}{P(F)} \\[2mm]
&= \frac{P(E \cap F)}{P(E \cap F) + P(E^c \cap F)} \\[2mm]
&= \frac{P(E) \cdot P(F \mid E)}{P(E) \cdot P(F \mid E) + P(E^c) \cdot P(F \mid E^c)} \\[2mm]
&= \frac{\text{Product of the probabilities along the path through } E \text{ and } F}{\text{Sum of the products of the probabilities along each path ending at } F}
\end{aligned}
$$

which is a special case of a result known as Bayes' theorem.

In general, Bayes' theorem is given as follows:

Let $E_1, E_2, \ldots, E_N$ be a partition of a sample space S and let F be an event of the experiment such that $P(F) \neq 0$. Then $P(E_i \mid F)$ (where $1 \leq i \leq N$) is given by

$$P(E_i \mid F) = \frac{P(E_i) \cdot P(F \mid E_i)}{P(E_1) \cdot P(F \mid E_1) + P(E_2) \cdot P(F \mid E_2) + \cdots + P(E_N) \cdot P(F \mid E_N)}$$

Example 7.6   A company gives a test to 120 salesmen, 80 with good sales records and 40 with poor sales records. Of the good salesmen, 65% pass the test, but only 20% of the poor salesmen pass the test.

a. What is the probability that a salesman passes the test?
b. If a salesman passes the test, what is the probability he is a poor salesman?
c. If a good salesman takes the test, what is the probability that he will not pass the test?

Solution
Let G denote the event that a good salesman takes the test, G^c denote the event that a poor salesman takes the test, F denote the event that a salesman fails the test, and F^c denote the event that a salesman does not fail (passes) the test. Using the given information, we can draw a tree diagram illustrating the situation (Figure 7.6).

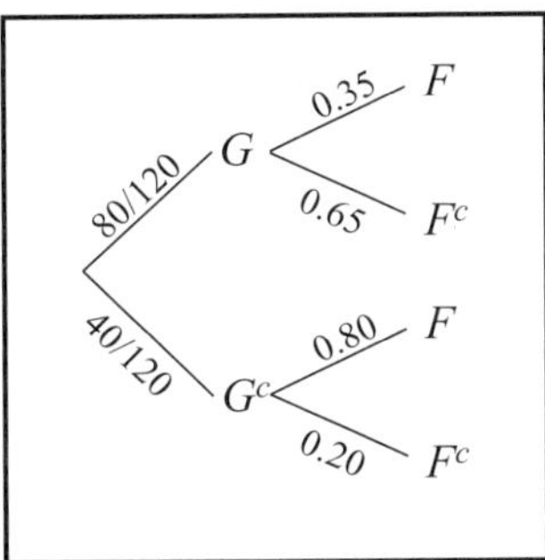

Figure 7.6   Company Sales Test Results

a. We are looking for $P(F^c)$. There are two disjoint ways to have this occur (two paths that end at F^c) and therefore,

$$P(F^c) = P(G \cap F^c) + P(G^c \cap F^c)$$
$$= \left(\frac{80}{120}\right)(0.65) + \left(\frac{40}{120}\right)(0.20)$$
$$= 0.50$$

b. We are looking for $P(G^c \mid F^c)$, which is not given on our tree. Therefore, using the conditional probability identity and the probability found in part (a), we find

$$P(G^c \mid F^c) = \frac{P(G^c \cap F^c)}{P(F^c)}$$
$$= \frac{\left(\frac{40}{120}\right)(0.20)}{0.50}$$
$$= \frac{2}{15} \approx 0.1333$$

c. We are looking for $P(F \mid G)$ which is a value on our tree and is equal to 0.35. ❖

Note: *You should now complete Activity C in the Conditional Probability Module.*

FURTHER EXPLORATIONS

For further explorations in the area of conditional probability, please visit the URL

http://www.finitemathtutor.com/explore/chapter7/

EXERCISES

Exercise 7.1   A bookshelf has 15 books on it. Nine books are blue and 6 are red. There are 3 red science books, 7 blue math books, and 5 psychology books. What is the probability that a psychology book is blue?

Use the following scenario for Exercises 7.2 and 7.3:

At a football game 85% of the spectators buy something to drink. Only 10% of the spectators who do not buy a drink buy a hot dog, whereas 25% of the spectators who buy a drink buy a hot dog.

Exercise 7.2   What is the probability that a spectator buys a hot dog?

Exercise 7.3   What is the probability that a spectator who bought a hot dog did not buy a drink?

Exercise 7.4   Two fair six-sided dice are rolled. What is the probability that the sum is greater than 10, given that at least one 5 is showing?

Exercise 7.5   Susie and Karen each have an identical shirt. Susie has a 10% chance of wearing the shirt to school on Monday and Karen has a 25% chance of wearing the shirt to school on Monday. If their decisions of what to wear to school on Monday are independent of one another, what is the probability that at least one of the girls will wear the shirt to school on Monday?

Exercise 7.6   Two cards are drawn in succession from a single well-shuffled deck of 52 cards. You are told that the second card drawn was a diamond. What is the probability that the first card drawn was a heart?

Use the following scenario for Exercises 7.7 and 7.8:

A bag has 5 yellow and 3 green balls. A basket has 1 yellow and 4 green balls. A ball is drawn at random from the bag and put into the basket. A ball is then drawn from the basket.

Exercise 7.7   Given that the ball chosen from the bag is yellow, what is the probability that a yellow ball is chosen from the basket?

Exercise 7.8   Given that a green ball is chosen from the basket, what is the probability that a green ball was chosen from the bag?

Use the following scenario for Exercises 7.9 and 7.10:

A store purchases T-shirts from 3 different companies. Twenty-five percent of the T-shirts are from company A and 50% of these T-shirts have pockets. Forty percent of the T-shirts are from company B and 10% of these shirts have pockets. The remaining T-shirts are from company C and none of these shirts have pockets.

Exercise 7.9   What is the probability that a T-shirt at this store has a pocket?

Exercise 7.10   What is the probability that a T-shirt at this store without a pocket comes from company B?

SAMPLE QUIZ

Question 7.1   A plate has 40 cookies on it and 10 of them are broken. There are 12 chocolate chip cookies (3 broken), 8 sugar cookies (4 broken), 15 oatmeal raisin cookies (1 broken) and 5 ginger snaps. What is the probability that a broken cookie is a ginger snap?

Use the following scenario to answer Questions 7.2 and 7.3:

At a local school 40% of the children buy lunch and 60% bring lunch. Forty percent of the children who buy lunch eat their entire lunch and 90% of the children who bring lunch eat their entire lunch.

Question 7.2   What is the probability that a child eats his entire lunch?

Question 7.3   Given that a child ate his entire lunch, what is the probability that the child brought the lunch from home?

Question 7.4   Two fair six-sided dice are rolled. What is the probability that the sum is less than 5, given that at least one 2 is showing?

Question 7.5   The chance of rain on Thursday is 30% and the chance of rain on Friday is 40%. If these rain chances are independent, what is the probability that it will rain at least once on these days?

Question 7.6   Two cards are drawn in succession without replacement from a standard deck of 52 cards. The second card is a jack. What is the probability that the first card drawn was a queen?

Use the following scenario to answer Questions 7.7 and 7.8:

Urn A contains 3 purple and 4 pink marbles. Urn B contains 2 purple and 3 pink marbles. A marble is drawn at random from urn A and placed into urn B. A marble is then selected from urn B.

Question 7.7   Given that the marble chosen from urn A was pink, what is the probability that the marble chosen from urn B is pink?

Question 7.8   Given that the marble chosen from urn B is purple, what is the probability that the marble chosen from urn A was pink?

Use the following scenario to answer Questions 7.9 and 7.10:

An appliance store buys dishwashers from 4 major companies. Thirty percent of the dishwashers are from company A and 10% of these dishwashers have electronic controls. Twenty percent of the dishwashers are from company B and 80% of these dishwashers have electronic controls. Forty percent of the dishwashers come from company C and 5% of these dishwashers have electronic controls. The remaining 10% of the dishwashers come from company D and none of them have electronic controls.

Question 7.9   What is the probability that a dishwasher at this store has electronic controls?

Question 7.10   What is the probability that a dishwasher without electronic controls comes from company A?

Statistics

8

When analyzing large amounts of data, we would like to answer the following questions:

- How do we describe the data set?
- How do we compare different data sets?
- Can we characterize important properties of the data set with just a few numbers?

Our goal in this chapter is to develop a **quantitative description** of large data sets. Interpretation of such descriptions is a more advanced topic that will be taken up elsewhere. Examples of quantitative descriptions of data include the following:

- The **minimum** data value
- The **maximum** data value
- The **range** of the data values
- The **average** (including the mean and median) of the data values
- The **mode** of the data values

8.1 CHARACTERISTICS OF A DATA SET

If we have a large set, S, consisting of single real numbers, we can *sort* the data, either lowest value to highest value, or highest to lowest. Clearly we can find the minimum or maximum of the data by selecting the first and last members of the sorted set.

For the set S, the **minimum of the set**, x, has the property that

$$x \leq y, \text{ for every } y \in S$$

The **maximum of the set**, z, has the property that

$$z \geq y, \text{ for every } y \in S$$

The **range of a set** is the difference $z - x$ if the data consists of real numbers. If the data consists of integers, the range is the difference $z - x + 1$.

Example 8.1   Given the data set $A = \{8, 4, 2, 15, 7, 2, 11\}$, find the range of the data.

Solution

Notice that these values are integers and therefore we will use the formula $z - x + 1$. The lowest value in the list is 2, so this is the minimum value, x. The highest value in the list is 15, which is therefore the maximum value, z. The range is $15 - 2 + 1 = 14$.   ❖

We can also count the number of times a particular value occurs. Making a table of frequency of occurrence versus value is called a **simple frequency distribution**. This gives some insight into the data. In particular, it tells us the mode, or most commonly occurring data value. The mode is a very poor description of the data since *any* data value may be a mode (from the minimum value to the maximum value). Even worse, if every value occurs just once there is *no mode*. If more than one value occurs the largest number of times, then each of those values is a mode. Therefore a data set can have no mode, one mode or more than one mode!

Example 8.2   What is the mode of data set A in the previous example?

Solution

The value 2 appears the most often (twice), so the mode is 2.   

Example 8.3   Given the data set B in the frequency table below, find the range and mode of B.

X	12	13	14	15	16	17
Frequency	2	5	1	6	6	3

Solution

The lowest x value is 12 and the highest is 17. The range is then $17 - 12 + 1 = 6$. The frequency is how often a data value occurs, so the x value that has the highest frequency is the mode. The highest frequency is 6 and this occurs for the x values of 15 and 16. The mode is, therefore, 15 and 16 (this is a bimodal distribution since there are two modes). ❖

 **Note:** *You should now complete Activity A in the Statistics Module.*

Large unordered data sets are extremely difficult to analyze without some additional tools. First we make analysis of the data simpler by grouping data into smaller subsets.

8.2   THE HISTOGRAM

If we divide the *range* of the data into equally sized intervals, and then collect data belonging to the same interval into **categories**, or subsets of the range, we arrive at a **grouped frequency distribution**. The number of groups or categories is rather arbitrary, but too few or too many categories is not very useful. Some general guidelines have been developed:

- Frequency distribution tables should have between 10 and 20 intervals.
- Interval sizes should be simple multiples of 2, 5, 10, 25, 50, 100, and so on.
- Intervals begin with the largest number that is divisible by the interval width and below the minimum value in the data set.
- All intervals are of equal width.

Example 8.4   The scores from 100 exams are data set C, shown below.

$C = \{$90, 89, 88, 71, 90, 72, 92, 61, 63, 64, 90, 77, 92, 86, 98, 52, 38, 79, 72, 70, 60, 42,
78, 79, 62, 66, 75, 56, 53, 76, 64, 77, 62, 80, 68, 55, 61, 80, 78, 61, 68, 87, 80, 91,
67, 80, 68, 45, 66, 97, 97, 49, 81, 60, 71, 82, 87, 82, 73, 73, 68, 76, 73, 80, 51, 85,
84, 81, 63, 66, 81, 74, 67, 76, 60, 81, 91, 75, 87, 45, 82, 78, 97, 62, 82,63, 72, 59, 48,
68, 97, 76, 96, 81, 70, 72, 53, 91, 72, 74$\}$

a. What is the range of values in data set C?
b. Make a frequency distribution table with 10 intervals.
c. Graph a frequency histogram for data set C.

Solution

a. The lowest value in the list is 38 and the highest is 98. The range is $98 - 38 + 1 = 61$.
b. Having ten intervals for a range of 61 means that the intervals should be 6 or 7 units wide. Since an interval is characterized by its midpoint, we shall choose 7 units wide so the midpoints will be integers. There is some flexibility in the starting point of the distribution; we will choose to start at 33.

Range	33–39	40–46	47–53	54–60	61–67	68–74	75–81	82–88	8995	96–102
Midpoint	36	43	50	57	64	71	78	85	92	99
Number	1	3	6	6	16	19	23	11	9	6

c. The histogram, as generated by the Statistics Applet, is shown in Figure 8.1. The number on the top of each column is the number of data values in that interval. The appearance of the histogram can change remarkably depending on the number of intervals. Two more histograms for the same data, but with different interval widths, are shown in Figures 8.2 and 8.3.

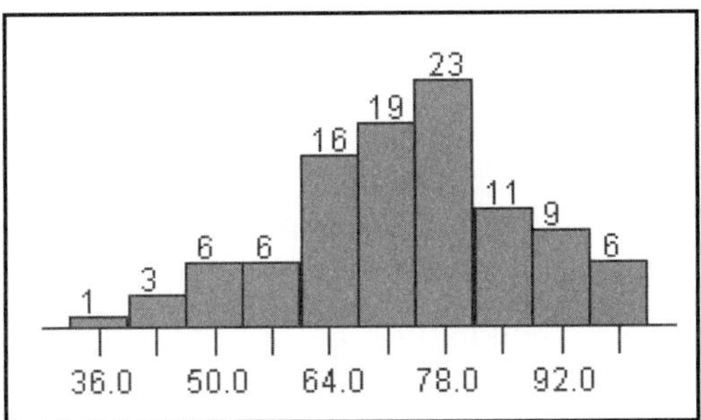

Figure 8.1   Histogram of Data Set *C* with Ten Categories

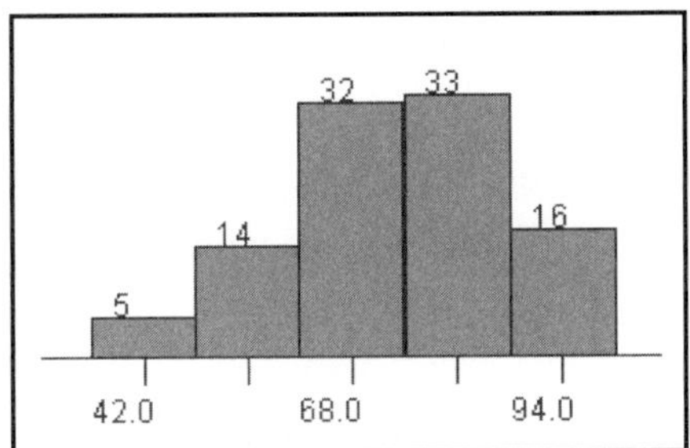

Figure 8.2   Histogram of Data Set *C* with Five Categories

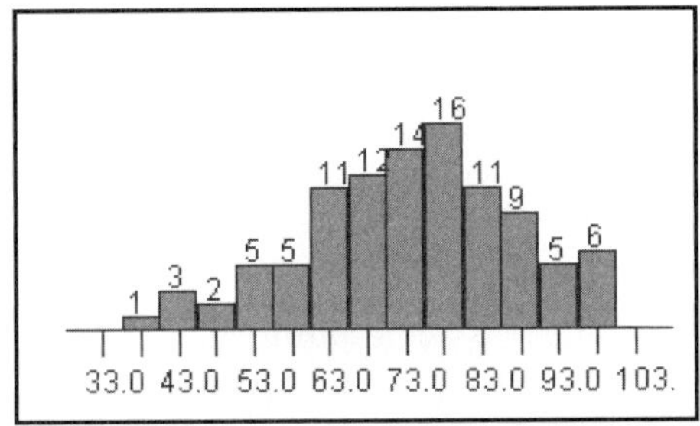

Figure 8.3   Histogram of Data Set *C* with Fifteen Categories

Clearly, the more intervals you have, the more information that can be seen — up to a point.   ❖

Note: *You should now complete Activity B in the Statistics Module.*

8.3   THE MEDIAN OF A DISTRIBUTION

We are often interested in where the center of the distribution is, that is, a **measure of central tendency** of the data. It should be somewhere near the "middle" of the histogram, but can we be more precise? One very precise measure is a quantity known as the **median**. If you rank order the data from lowest to highest, the median is the value of the data point for which half of the data is less, and half of the data is more.

For example, if you have 11 unique data points, 5 data points will lie below the 6th data point, and 5 points will lie above. In the case of an even number of data points, we average the two points closest to the middle, so in the case of 10 data points, we average the values of the 5th and the 6th data points. This is unambiguously defined, and gives a very robust measure of central tendency. Since 50 percent of the data lies below the median, it is also called the 50th percentile. The 25th percentile is the value that has one-fourth of the data below it and three-fourths of the data above it. This is also called the first **quartile** or Q_1. The third quartile, Q_3, is that value that separates the top one-fourth of the data from the bottom three-fourths of the data.

Example 8.5   Find the median and quartiles for data sets A, B, and C (given in Examples 8.1, 8.3, and 8.4).

Solution

For data set A, start by putting the data in order from lowest to highest:

$$2, 2, 4, 7, 8, 11, 15$$

With 7 values, the midpoint occurs at the 4th value (3 values are above this and 3 values are below this). The median is, therefore, 7. The first quartile will be at the 2nd data point, 2, and the third quartile will be at the 6th data point, 11.

For data set B, we need to expand the frequency table:

$$12, 12, 13, 13, 13, 13, 13, 14, 15, 15, 15, 15, 15, 15, 16, 16, 16, 16, 16, 16, 17, 17, 17$$

There are 23 data points, so the median will be the 12th data point, 15. The first quartile is $Q_1 = 13$ and the third quartile is $Q_3 = 16$.

For data set C, it is best to use a graphing calculator or computer to sort the list and find the quartiles. When the cutoff for a quartile or median comes between two data points, we must interpolate between the two values. We find that $Q_1 = 63.5$, median $= 74$, and $Q_3 = 81.75$. The locations of the 25th, 50th, and 75th percentiles on the histogram are shown in Figure 8.4 from the Statistics Applet.

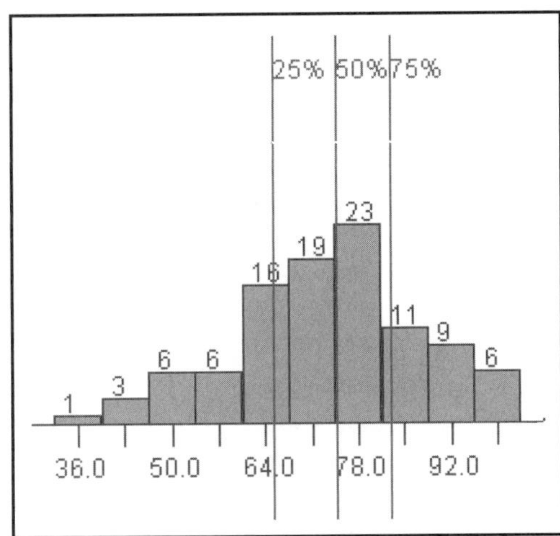

Figure 8.4 Percentiles

Although it may seem so, the median does not necessarily lie on an interval boundary. It is also apparent that the 25th, 50th, and 75th percentiles may be quite close together, relative to the range of the data.

8.4   THE MEAN OF A DISTRIBUTION

The most robust and consistent estimator of central tendency is called the mean. It is closely related to the concept of the average of a set of numbers and the **expected value** of a random variable. Suppose we have the following set of data

$$S = \{1, 1, 2, 3, 5, 5, 5, 7, 10, 10\}$$

We can easily calculate the mean as

$$\mu = \frac{1+1+2+3+5+5+5+7+10+10}{10}$$

$$= 1 \cdot \left(\frac{2}{10}\right) + 2 \cdot \left(\frac{1}{10}\right) + 3 \cdot \left(\frac{1}{10}\right) + 5 \cdot \left(\frac{3}{10}\right) + 7 \cdot \left(\frac{1}{10}\right) + 10 \cdot \left(\frac{2}{10}\right)$$

We obtain the last form of the equation by taking the frequency of occurrence of each data point and dividing by the total number of data points. This gives the relative frequency of each data point and is equal to the theoretical probability of a particular data point occurring when chosen at random from the entire sample of data points. We then multiply the value of each different data point by this theoretical probability.

In general, this leads to the formula for the mean of a data set with M values

$$\boxed{\mu = x_1 p_1 + x_2 p_2 + \cdots + x_M p_M}$$

x_i is the value of the ith data point and p_i is the theoretical probability of this value occurring.

 **Note:** *You should now complete Activities C – E in the Statistics Module.*

8.5   MEASURES OF SPREAD

Now that we have identified several measures of central tendency, we look at another quantitative concept — measures of the **spread** or **variability of a distribution**. Recall that we defined the range of a distribution as the difference between the maximum and minimum values of the distribution. This gives us an upper bound on any possible definition of the spread of a distribution: All such measures must be less than or equal to the range.

Another measure of variability is the **interquartile range** Q, that is, the difference between the 75th and the 25th percentile. It is the range in which the middle 50% of the data lies. It does not have to be symmetric about the median. The interquartile range also helps us define the concept of an **outlier**. An outlier is a point that lies more than $1.5 \times Q$ below (or above) the 25th (or 75th) quartile. A box plot can help illustrate this point.

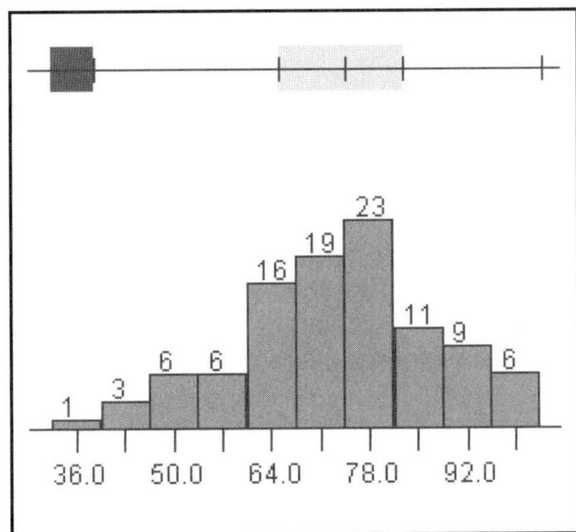

Figure 8.5   Box Plot

On the line at the top of Figure 8.5, the middle shaded area indicates the interquartile range, and the left shaded range is data which is less than below the 25th percentile.

Another possibility is to take the mean of the *distance* of each data point away from the mean of the distribution. Letting μ denote the mean of the distribution, and writing the distance of the ith point from the mean as $x_i - \mu$, then the average of the distances is equal to

$$\sum_{i=1}^{M} |x_i - \mu| p_i$$

The **summation notation** is defined in the glossary. This is not as useful in practice as two other quantities, the **variance**

$$\sigma^2 = \sum_{i=1}^{M} |x_i - \mu|^2 \, p_i$$

and the **standard deviation**

$$\sigma = \sqrt{\sum_{i=1}^{M} |x_i - \mu|^2 \, p_i}$$

which is the **root mean square** distance from the mean. A root mean square is the square root of the average of the squares. The standard deviation is by far the most useful, robust, and common measure of the spread of data.

Example 8.6   Find the variance and the standard deviation for data set A.

Solution

Data set A is $\{8, 4, 2, 15, 7, 2, 11\}$ and the mean is 7. We can use a table to organize the numbers needed to find the variance.

x	$x - \mu$	p	$(x - \mu)^2 p$
8	1	$\frac{1}{7}$	$\frac{1}{7}$
4	-3	$\frac{1}{7}$	$\frac{9}{7}$
2	-5	$\frac{1}{7}$	$\frac{25}{7}$
15	8	$\frac{1}{7}$	$\frac{64}{7}$
7	0	$\frac{1}{7}$	0
2	5	$\frac{1}{7}$	$\frac{25}{7}$
11	4	$\frac{1}{7}$	$\frac{16}{7}$
Total			$\frac{140}{7} = 20$

The variance is 20 and the standard deviation is $\sigma = \sqrt{20} \approx 4.47$.

The Statistics Applet will also find the standard deviation for a data sample (Figure 8.6). The standard deviation for a sample is a little larger than the standard deviation for a probability distribution.

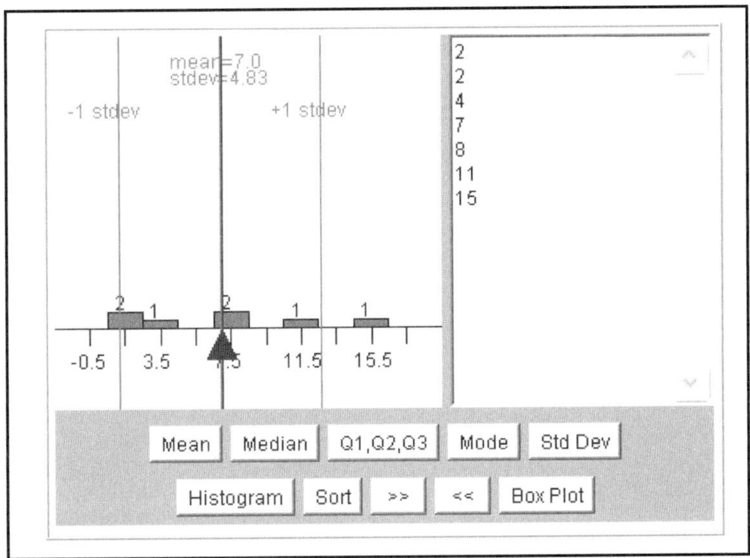

Figure 8.6   Standard Deviation Using the Statistics Applet   

 **Note:** *You should now complete Activities F and G in the Statistics Module.*

 # FURTHER EXPLORATIONS

For further explorations in the area of Statistics please visit the URL

http://www.finitemathtutor.com/explore/chapter8/

EXERCISES

Exercises 8.1–8.5 use the following data: $D = \{-2, 0, 49, 1, 3, 6, -1, 0, 1\}$.

Exercise 8.1   What is the range of data set D?

Exercise 8.2   What is the mode of data set D?

Exercise 8.3   What is the mean of data set D?

Exercise 8.4   What is the median of data set D?

Exercise 8.5   Which measure of central tendency is least representative of data set D?

Exercises 8.6–8.8 use the following histograms:

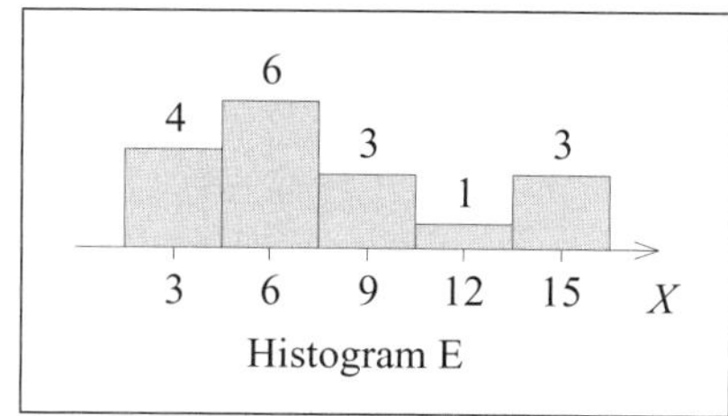

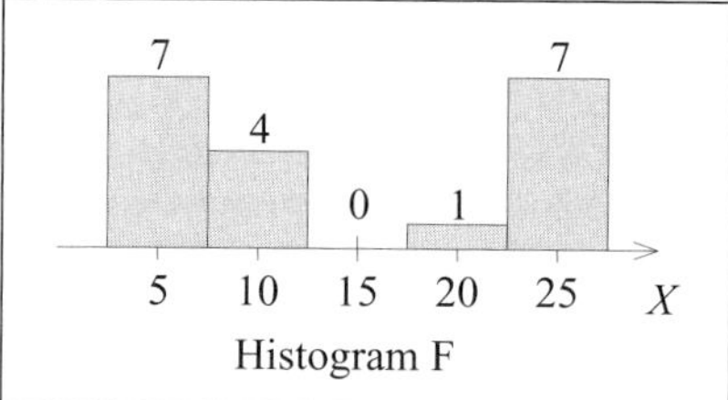

Exercise 8.6   How wide are the intervals in Histogram E and Histogram F?

Exercise 8.7   What is the mode for Histogram E and Histogram F?

Exercise 8.8   Which histogram has the larger standard deviation?

Exercises 8.9 and 8.10 use the following frequency distribution:

X	10	11	12	13	14	15	16
Frequency	2	6	8	5	0	2	1

Exercise 8.9   What is the mode?

Exercise 8.10   What are the mean and median?

SAMPLE QUIZ

Questions 8.1–8.5 use the following data set: $M = \{5, 3, 100, 5, 11, 7, 9\}$.

Question 8.1   What is the range of data set M?

Question 8.2   What is the mode of data set M?

Question 8.3   What is the mean of data set M?

Question 8.4   What is the median of data set M?

Question 8.5   Which measure of central tendency is least representative of data set M?

Question 8.6   A grouped data set has midpoints at 12, 16, 20, and 24. How wide are the intervals?

Questions 8.7 and 8.8 use the following histograms:

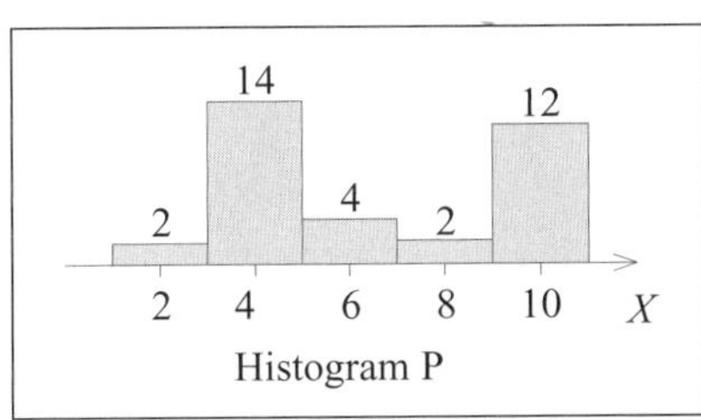

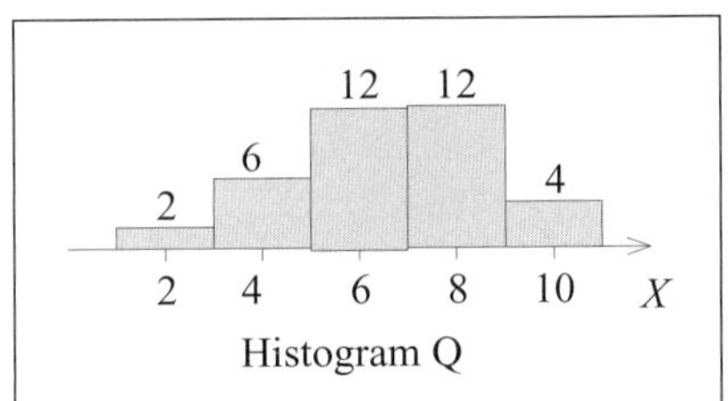

Question 8.7   What is the mode of Histogram P and Histogram Q?

Question 8.8   Which histogram has the largest variance?

Questions 8.9 and 8.10:

Small bags of sour worms candy are opened and the number of blue worms in each package are counted. The following are the results of this experiment:

Number of Blue Worms	0	1	2	3	4	5	6
Number of Bags	2	7	12	7	10	0	1

Question 8.9   What is the mode for this experiment?

Question 8.10   What is the mean?

Probability Distributions

For any event, there is a probability of occurrence. This probability ranges from 0 (never occurs) to 1 (absolute certainty). If we look at a set of events occurring, with different probabilities, the result is a **probability distribution**. There are many types of probability distributions; we consider some of the most common in this chapter.

9.1 PROBABILITY AND RANDOM VARIABLES

An **experiment** is any activity with an outcome that can be observed. Examples of experiments include dropping an object and measuring its velocity or time to impact, finding the rate of reaction when two chemicals are mixed together, or flipping a coin 100 times and counting the number of heads that occur. A single outcome of an experiment is sometimes called a **sample point**. The set of all possible outcomes of an experiment is called the **sample space**. An **event** is a subset of the sample space, consisting of one or more of the outcomes of an experiment. If a **finite number of outcomes** are possible, then the probabilities are likely to be non-zero. If there is an **infinite number of outcomes**, the probability of a single specific outcome is likely near zero!

If a probability distribution has a finite number of outcomes associated with the sample space, it is called a **discrete distribution**. If it takes on all values in a *range* of probabilities, it is called a **continuous distribution**. The outcome of an experiment can be assigned a numerical value. We call this value a **random variable**. It is random since we do not know what outcome will occur until we do the experiment. It is a variable since it can take on different values. When our random variable is discrete, we can write a probability distribution table to summarize the probabilities of each of the possible outcomes of our experiment. When our random variable is continuous, the probability distribution can be represented by a function.

Example 9.1   Suppose S is the sample space associated with rolling a six-sided die and observing the outcome. What is the random variable associated with this experiment? What are the possible values of the random variable? Write a probability distribution table for this experiment.

Solution

The simplest possible random variable is just counting the number of dots showing on the upward face (X itself!). There are six possible outcomes and if the die is fair, the probabilities are equally likely, as shown in Table 9.1.

X	$P(X)$
1	$\frac{1}{6}$
2	$\frac{1}{6}$
3	$\frac{1}{6}$
4	$\frac{1}{6}$
5	$\frac{1}{6}$
6	$\frac{1}{6}$

Table 9.1   Probability Distribution Table for a Single Die   

However, suppose we throw *two* dice at the same time. If X is the random variable associated with counting the dots on the first die, and Y is associated with counting the dots on the second die, then we could define a new random variable, $Z = X + Y$, which counts the total number of dots on the throw. We could also define random variables $Z = X \cdot Y$ (the product), $Z = X / Y$ (the quotient), and so on. There is an infinite number of random variables.

In Example 9.1, we saw the first example of a probability distribution. It is an important special case, where all the probabilities are equal, and is called a uniform probability distribution. We look at this case in more detail in the next section.

Where there is no danger of confusion, we refer to a random variable with a particular probability distribution, for example, a **uniformly distributed random variable**.

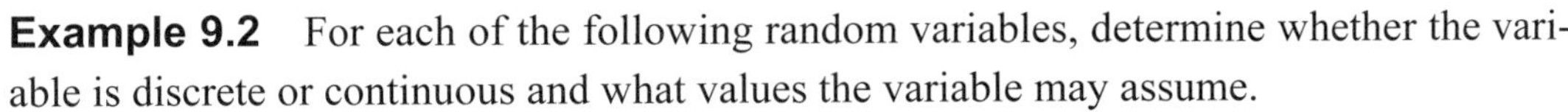

Example 9.2   For each of the following random variables, determine whether the variable is discrete or continuous and what values the variable may assume.

a. X = the number of girls in a class of 30 students
b. X = the number of hours you study in a two-day period
c. X = the number of times you flip a coin until it lands on heads

Solution

a. X is discrete where X takes on values in the set $\{0, 1, 2, \ldots, 30\}$.
b. X is continuous where X takes on values in the set $\{t \mid 0 \leq t \leq 48\}$ where t is any real number. Notice that this set of values is different than saying X takes on values in the set $\{0, 1, 2, \ldots, 48\}$, because this set does not include all of the possible study hours, for example 1.1 hours or 1.01 hours.
c. X is discrete where X takes on values in the set $\{1, 2, \ldots\}$. Notice that there is no upper limit on the value that X may assume since there is no way to know how many tosses it will take for the coin to land on heads, but you do know that the value X assumes will be an integer value.   ❖

9.2   UNIFORM DISTRIBUTION

If the probabilities of all of the outcomes of an experiment are the same, we call the distribution of probabilities a **uniform probability distribution**. A simple example of such a distribution is the toss of a single coin, with heads on one side and tails on the other. Unless the coin has been altered by some means, the probability of getting a head is $\frac{1}{2}$, the same as the probability of getting a tail. Let the random variable X be the number of heads (which will be graphed along the x-axis). With the probabilities graphed on the y-axis, we can get a probability histogram as shown in Figure 9.1.

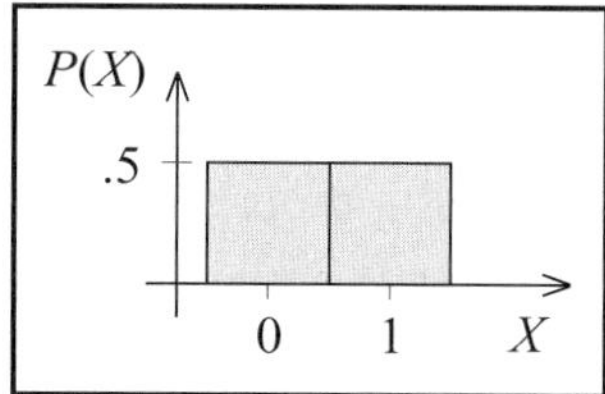

Figure 9.1   Probability Distribution of Coin Toss

Example 9.3   A game spinner is divided into 12 equal sections as shown in Figure 9.2. If each section has the same probability of occurring after the arrow is spun, find the probability distribution for this game spinner.

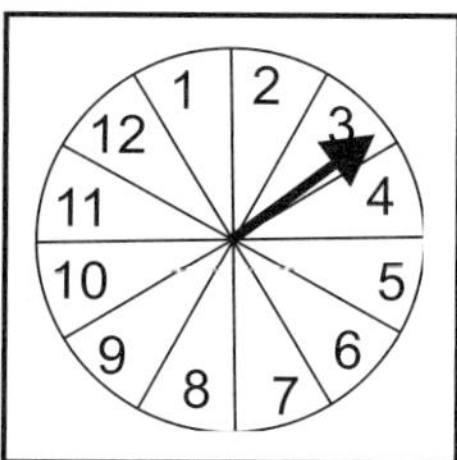

Figure 9.2   A Game Spinner

Solution

Each of the outcomes has a $\frac{1}{12}$ probability and the probability distribution is shown in Figure 9.3.

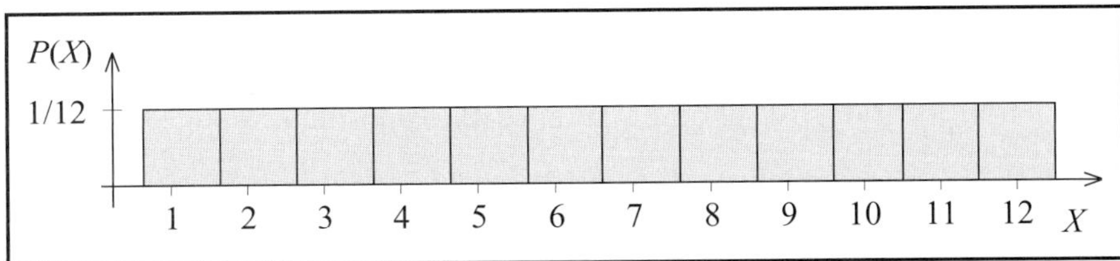

Figure 9.3   Game Spinner Probability Distribution   ❖

.3   BERNOULLI DISTRIBUTION

As a special case of an experiment, consider the following:

A fair coin is tossed 100 times in succession. After each toss, the outcome is recorded — either a head or a tail.

We note several things about this experiment. First of all, we perform only a fixed number of trials, in this case 100. There are only two possible outcomes for each trial. The probability of each outcome remains constant. Each trial is independent — the probability of a head or tail occurring does not depend on whether a head or tail occurred on the previous trial. Even if 99 heads in a row occur, if the coin is fair, the probability of getting a head on the 100th toss is still $\frac{1}{2}$.

This kind of experiment occurs so frequently that it is called a **binomial** (or Bernoulli) experiment. An experiment is *binomial* if

- The number of trials, n, is fixed.
- The only outcomes of each trial are success and failure.
- The trials are independent of each other and so the probability of success, p, is the same for each trial.

The terms *success* and *failure* are relatively arbitrary and refer to one or the other of the outcomes. If X is a binomial random variable, then the probability, $P(X = x)$, of x successes in the n trials is given by the binomial probability formula

$$P(X = x) = C(n,x) \cdot p^x \cdot (1-p)^{n-x}$$

where p is the probability of success in each trial and $C(n, x)$ is the number of ways that the x successes can occur in the n trials (see Chapter 6).

Example 9.4   A fair six-sided die is rolled 20 times. What is the probability that a five is rolled 3 times?

Solution

This is a binomial experiment with our random variable, X, denoting the number of fives rolled. We have $n = 20$, $p = \frac{1}{6}$, and $x = 3$. This leads to

$$P(X = 3) = C(20,3) \cdot \left(\frac{1}{6}\right)^3 \cdot \left(\frac{5}{6}\right)^{17}$$

$$\approx 0.2379$$

We can also use the Binomial Probability Calculator applet to find our probability by filling in the given values and then clicking on the compute button (Figure 9.4).

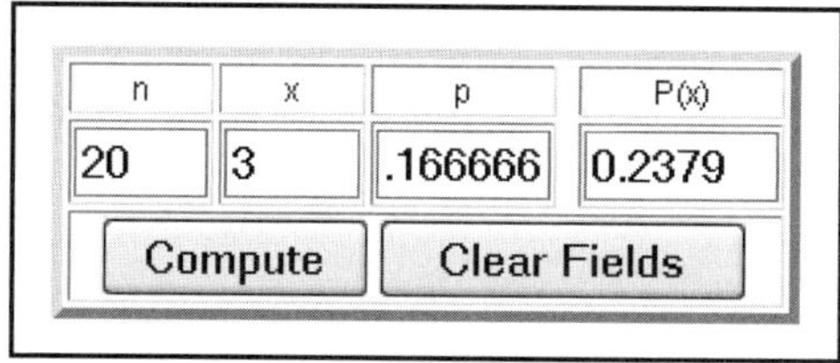

Figure 9.4   Binomial Probability Calculator

In Example 9.4, the random variable X is a discrete random variable that can take on the values 0, 1, 2, 3, ..., 20. In other words, it is possible to roll 0 fives, 1 five, on up to 20 fives. We can calculate the probability that X attains each of these values, as we did for $X = 3$, and find the probability distribution for X (Figure 9.5).

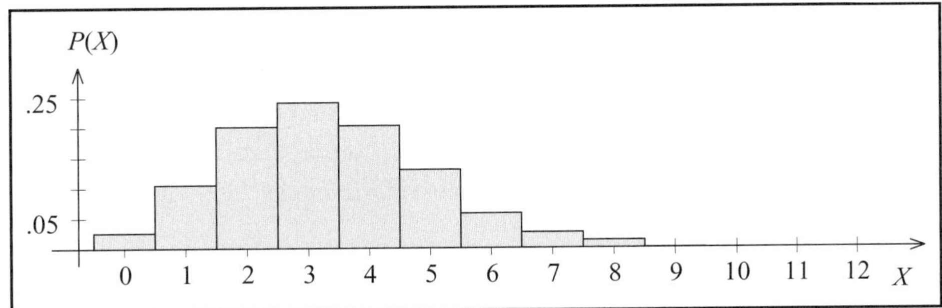

Figure 9.5   Probability Distribution of the Number of Fives in 20 Rolls

You can use the probability distribution of a discrete random variable to find the probability that the random variable lies in a certain range. For instance, in the previous example, calculating the probability $P(2 \leq X \leq 5)$ amounts to finding $P(X = 2) + P(X = 3) + P(X = 4) + P(X = 5)$. Moreover, because of how a histogram is constructed, this amounts to adding the areas of the rectangles centered about 2, 3, 4, and 5 in Figure 9.5.

The mean and standard deviation of a **binomial distribution** can be found using the binomial statistics formulas:

$$\mu = np \qquad \sigma = \sqrt{np(1-p)}$$

where n is the number of trials and p is the probability of success in a single trial.

 **Note:** *You should now complete Activity A in the Probability Distributions Module.*

9.4   THE NORMAL DISTRIBUTION

The **normal distribution** is a continuous probability distribution with a given mean (μ) and standard deviation σ. It is modeled by the function

$$P(x) = Ae^{-\frac{1}{2}(x-\mu)^2/\sigma^2}$$

where A is chosen to ensure that the total probability of the sample space $\{-\infty < X < \infty\}$ is equal to one. The distribution is centered about the mean and the standard deviation determines the width of the distribution. An example of a normal distribution is shown in Figure 9.6. One can see why normal distributions are referred to as "bell curves."

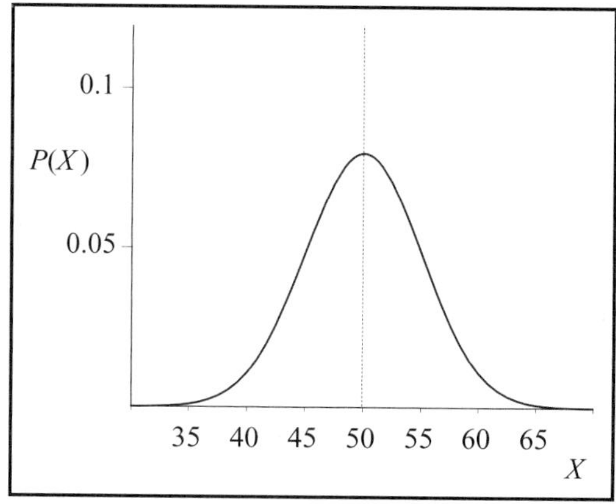

Figure 9.6   A Normal Distribution with $\mu = 0$ and $\sigma = 5$

The probability that an outcome of the experiment will be between the values X_1 and X_2 is equal to the area under the normal curve between X_1 and X_2. The area can be found using tables, a graphing calculator, or the Normal Distribution Applet. A **standard normal distribution** has $\mu = 0$ and $\sigma = 1$. Figure 9.7 shows a standard normal distribution with the area between $X_1 = -1$ and $X_2 = 1$ shaded using the Normal Distribution Applet. The probability that a value lies in this range is 0.6826.

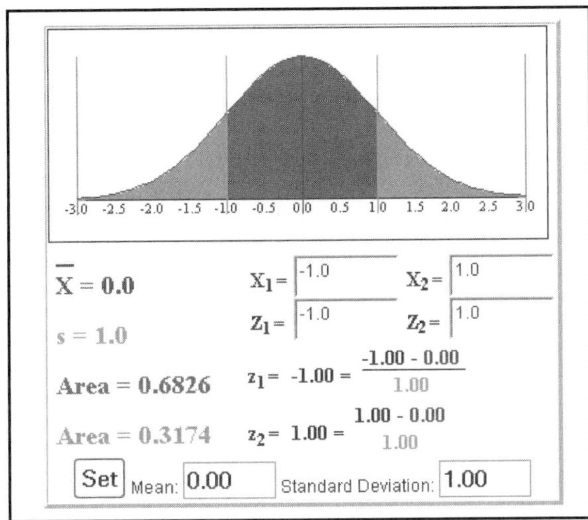

Figure 9.7   Standard Normal Distribution

A non-standard normal variable can always be converted to a **standard normal variable** Z by means of the transformation

$$Z = \frac{(X - \mu)}{\sigma}$$

This is useful, since tables of probabilities are usually given only for standard normal distributions. Alternatively, we can use the Normal Distribution Applet to find the probabilities of normal random variables.

Example 9.5   Suppose X is a normal random variable with a mean of 12 and a standard deviation of 0.2. What is the probability that a value of X will be between 11.7 and 12.3?

Solution
Using the Normal Distribution Applet, first set the mean and standard deviation at the bottom of the screen. Make sure you choose the "SET" button before proceeding. For this problem, $X_1 = 11.7$ and $X_2 = 12.3$, so enter these values into the applet, and the area between 11.7 and 12.3 gives the probability you are looking for (Figure 9.8).

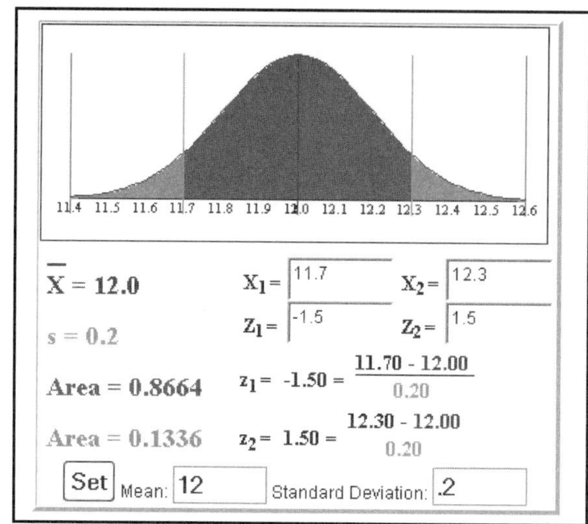

Figure 9.8   Normal Distribution Applet

Therefore, $P(11.7 \leq X \leq 12.3) = 0.8664$.   

 **Example 9.6**   The grades on a math test are normally distributed with a mean of 70 and a standard deviation of 8. What is the probability that a randomly selected test has a grade of 90 or higher?

Solution

First, it is important to notice that the grades, X, are *normally distributed*. This tells us that to calculate the probability we will use methods associated with a normal distribution. Using the Normal Distribution Applet, we will first set the mean = 70 and the standard deviation = 8. Because we are looking for $P(X \geq 90)$, we will let $X_1 = 90$ and $X_2 = 500$ (a number far enough away that it will act like infinity) and then allow the applet to calculate the probability (Figure 9.9).

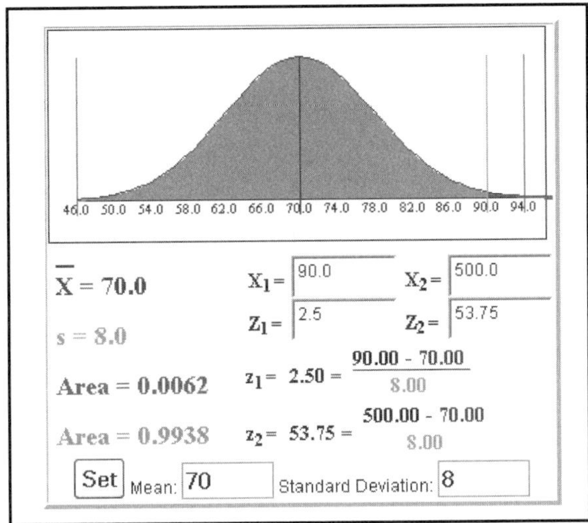

Figure 9.9   Probability of Test Grade

Therefore, $P(X \geq 90) = 0.0062$ ❖

Example 9.7   The instructor from Example 9.6 decides that she needs to "curve" the grades since less than 1% of the class received an A on the test. If the top 10% of the class should get an A, which of the following cutoffs should be used to decide the score needed for an A on the exam?

a. 72   b. 74   c. 76   d. 78   e. 80
f. 82   g. 84   h. 86   i. 88   j. 90

Solution
It is possible to "invert" the normal distribution using a graphing calculator [Chapter 9, Resources]. However, when given possible cutoffs, one can "guess and check" each option to see which has the desired area. As shown in Figure 9.10, answer **e** is the best choice since the area above $X = 80$ is about 10.5%.

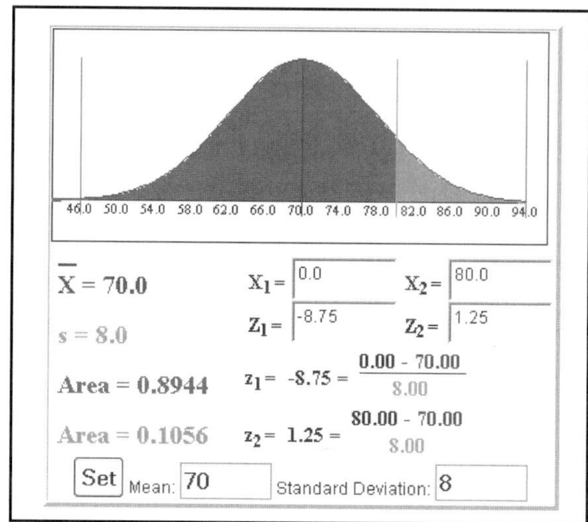

Figure 9.10   Finding Percentiles Using the Normal Applet   ❖

The total area under the normal curve is always 1. However, the larger the standard deviation, the more flattened the curve appears. Figures 9.11 through 9.13 show the effect of an increasing standard deviation.

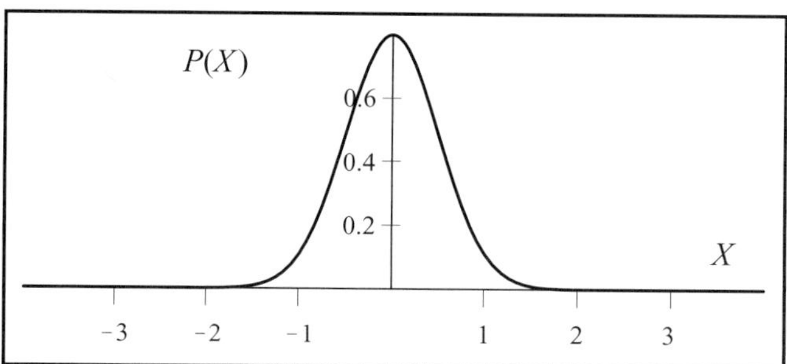

Figure 9.11   Normal Distribution with $\mu = 0$ and $\sigma = 0.5$

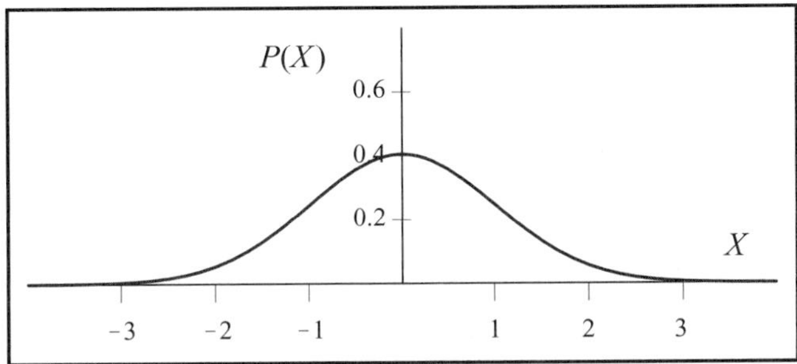

Figure 9.12 Normal Distribution with $\mu = 0$ and $\sigma = 1$ (The Standard Normal Curve)

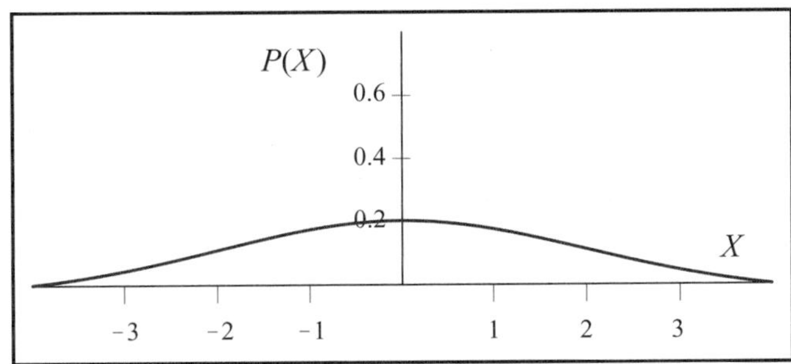

Figure 9.13 Normal Distribution with $\mu = 0$ and $\sigma = 2$

 **Note:** *You should now complete Activity B in the Probability Distributions Module.*

9.5 THE LAW OF LARGE NUMBERS

If one looks at distributions formed from large numbers of observations, for example the distribution of SAT scores for all high school students in the United States in a given year, one notices the distinctive "bell" shape of a normal distribution.

Tossing a coin 100 times, counting the number of heads, and then creating a histogram of the frequency distribution is another example. This can be accomplished quite rapidly through the Coin-Toss Applet. The sequence of Figures 9.14–9.17 illustrates 50,

100, 200, and 500 trials of the Coin-Toss Applet. One can see that although the initial distribution does not seem very much like a normal distribution, by the time one has reached 500 trials, the fit to a normal distribution is much closer. The approach to a normal distribution is called the **law of large numbers** (or **central limit theorem**). There will always be *some* difference between a binomial distribution and a normal distribution, but this difference becomes smaller as the number of trials gets larger.

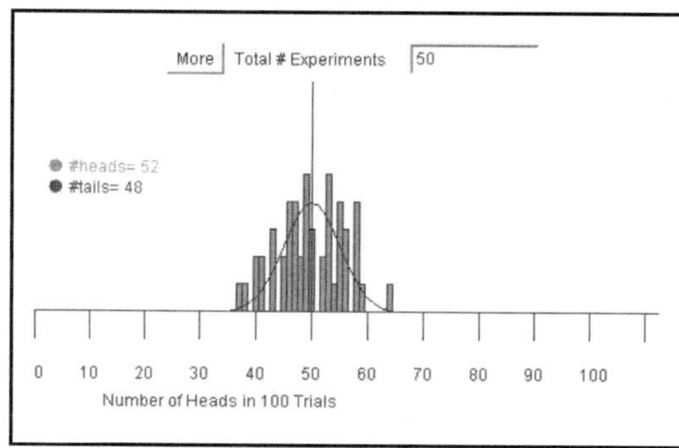

Figure 9.14   Fifty Trials

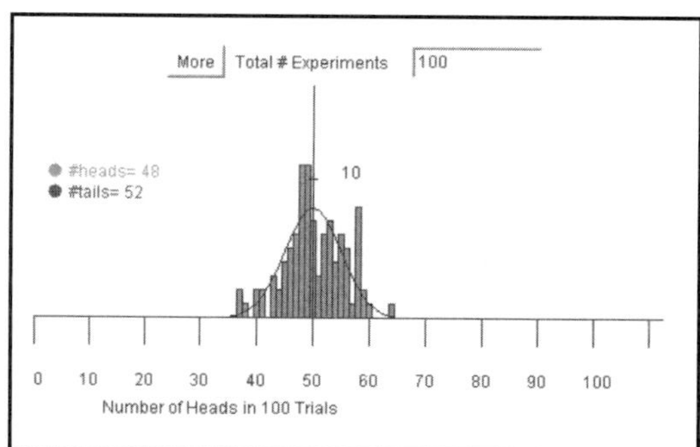

Figure 9.15   One Hundred Trials

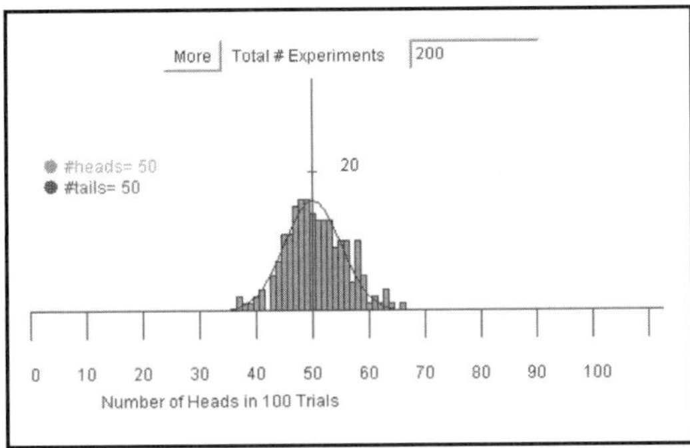

Figure 9.16   Two Hundred Trials

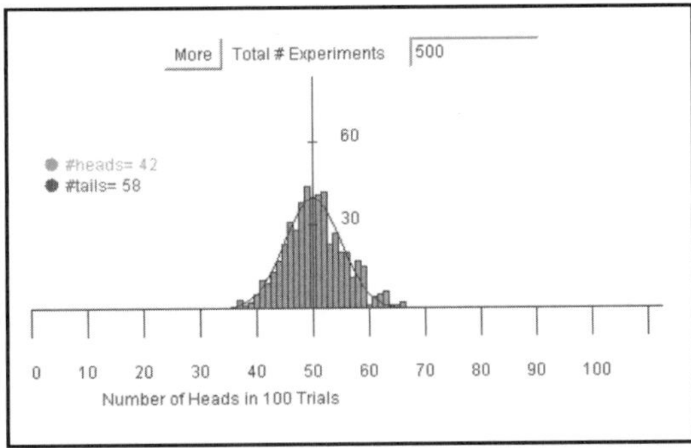

Figure 9.17   Five Hundred Trials

The law of large numbers will allow us to approximate a distribution — in this case, to approximate the binomial distribution with the normal curve. In Figure 9.5 we see the histogram for the binomial experiment in Example 9.4 where a die was rolled 20 times. The mean and standard deviation for this binomial experiment are

$$\mu = np = 20 \cdot \frac{1}{6} \approx 3.33 \qquad \sigma = \sqrt{np(1-p)} = \sqrt{20 \cdot \frac{1}{6} \cdot \frac{5}{6}} \approx 1.67$$

The normal curve with $\mu = 3.33$ and $\sigma = 1.67$ is superimposed on the histogram in Figure 9.18. The two distributions are quite close and will be even closer with larger values of n.

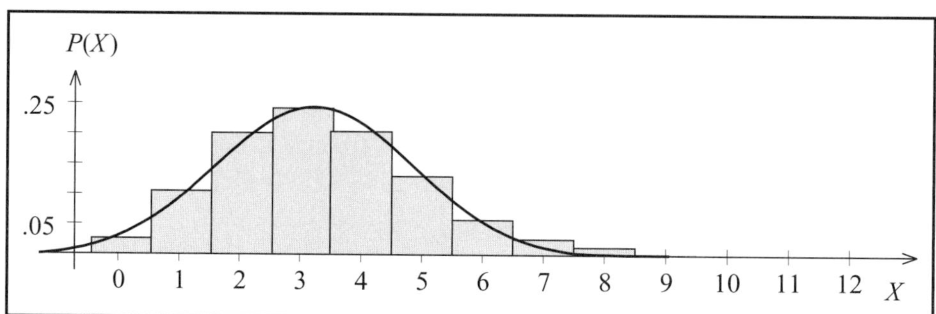

Figure 9.18   Normal Curve Approximation to the Binomial Distribution

This close agreement means that the normal curve can be used to approximate the binomial distribution. This is useful when there are many trials.

Example 9.8   A fair coin is tossed 1,000 times and the number of heads is counted.

a. What are the mean and standard deviation for the number of heads?
b. Use the normal curve to approximate the probability that there were at least 500 but no more than 525 heads tossed.

Solution

a. This is a binomial experiment and so

$$\mu = np = 1000 \times .5 = 500 \qquad \sigma = \sqrt{1000 \times .5 \times .5} \approx 15.81$$

b. The actual histogram for the binomial experiment has 1,001 rectangles centered on each of the possible values of the number of heads from 0 to 1,000. The exact probability is the area of the rectangles representing tossing exactly 500 heads, 501 heads, and so on up to exactly 525 heads. The rectangle on the histogram representing tossing exactly 500 heads starts at $X = 499.5$ and goes to $X = 500.5$. That means that the left endpoint for the area under the normal curve is 499.5. The rectangle for tossing exactly 525 heads starts at $X = 524.5$ and ends at $X = 525.5$. That means that the right endpoint for the area under the normal curve is 525.5. So we will find the probability is the area under the normal curve from 499.5 to 525.5 (Figure 9.19).

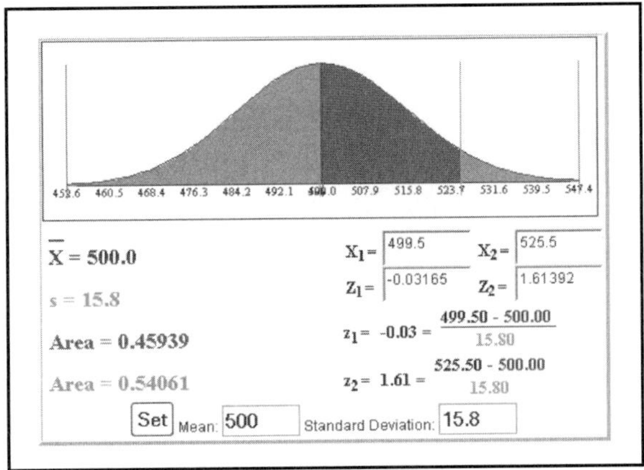

Figure 9.19 Normal Curve Approximation for 1,000 Coins

Therefore, the approximated probability is 0.4594.

Note: *You should now complete Activity C in the Probability Distributions Module.*

FURTHER EXPLORATIONS

For further explorations in the area of probability distributions please visit the URL

http://www.finitemathtutor.com/explore/chapter9/

EXERCISES

Exercise 9.1 A coin is tossed 6 times and the number of tails is counted. What is the probability that 4 tails are tossed?

Exercise 9.2 Three different experiments are done.

Let X_1 = the number of minutes you spend taking notes in math class.

Let X_2 = the number of cracked eggs in a dozen of eggs.

Let X_3 = the number of times the word "the" is written on this page.

Which of the random variables are discrete?

Exercise 9.3   A jar contains 50 jellybeans (10 pink, 20 yellow, 6 white, and 14 green). A jellybean is chosen, the color is noted, and then the jellybean is placed back into the jar. Let X be the number of jellybeans drawn until a white jellybean is chosen. What are the values that X may assume?

Exercise 9.4   A bag contains 10 balls, 5 red and 5 black. Four balls are drawn out one after another without replacement. Let X be the number of red balls picked. Is X a binomial random variable?

Exercise 9.5   Suppose X is a normal random variable with a mean of 40 and a standard deviation of 5. What is the probability that a value of X will be between 32 and 44?

Exercise 9.6   The net weight of a 10-ounce bag of Crunchy Chips has a mean of 9.8 ounces and a standard deviation of 0.2 ounces. Assuming the distribution of the net weight is normal, what is the probability that the net weight of a bag of chips is more than 3 standard deviations away from the mean?

Exercise 9.7   The scores on a math exam are normally distributed with a mean of 68 and a standard deviation of 13. If the top 5% of the students receive an A on the test and the next 15% receive a B, what is the lowest score a student may have and still obtain a B?

a. 57   **b.** 88   **c.** 80   **d.** 79   **e.** 75

Use the following scenario to answer Exercises 9.8–9.10:

During their most recent move, the Smith family moves a box containing 200 candles. En route to their final destination, each candle has a 3% chance that it will be broken.

Exercise 9.8   What is the mean and standard deviation for the number of broken candles?

Exercise 9.9   To find the probability that more than 8 candles are broken in the move, we can use the normal curve to approximate the binomial distribution. What are the endpoints that should be used?

Exercise 9.10 Using the normal curve approximation, what is the probability that when the candle box arrives there are more than 8 broken candles?

SAMPLE QUIZ

Question 9.1   A coin is tossed 8 times and the number of heads is counted. What is the probability that 5 heads are tossed?

Question 9.2   The mean weight of a box of cereal is 16 ounces with a standard deviation of 0.75 ounces. If the weight of the cereal box can be closely approximated with the normal distribution, what is the probability that a box of cereal weighs between 16 and 18 ounces?

Question 9.3   A pair of fair six-sided dice is rolled and the number of dots showing is counted. Let X be the number of rolls needed before a sum of 12 is found. What are the values that X may assume?

Question 9.4   A hand of 4 cards is dealt and the number of red cards is counted. Is this a binomial experiment?

Question 9.5   Forty corn seeds are planted for an experiment. Each seed has an 80% chance of sprouting. What are the mean and standard deviation for the number of seeds expected to sprout?

Use the following scenario to answer Questions 9.6 and 9.7:

A bag of raisins has a mean of 60 raisins and a standard deviation of 4 raisins.

Question 9.6   Using the normal curve approximation to the binomial distribution to find the probability that a bag of raisins has strictly between 60 and 66 raisins, what are the endpoints that should be used?

Question 9.7   Using the normal curve approximation to the binomial distribution, what is the probability that a bag of raisins has strictly between 60 and 66 raisins?

Question 9.8   A particular tennis player serves an ace 8% of the time. If this player serves 150 times, what is the probability that he will serve 20 aces?

Question 9.9   Suppose X is a normal random variable with a mean of 20 and a standard deviation of 4. What is the probability that a value of X is greater than 25?

Question 9.10   A bag contains 26 slips of paper. Each slip of paper has a different letter of the alphabet printed on it. One slip of paper is drawn from the bag without replacement until a vowel (A,E,I,O,U) is drawn. Let X be the number of slips of paper drawn. What are the values that X may assume?

Financial Applications

10

Mathematics is used constantly in everyday life. Banks and financial institutions use mathematical formulas to calculate interest on loans and on certificates of deposit. Traders on Wall Street use sophisticated mathematical models to price options and minimize risk. Insurance rates are determined using formulas based on mortality tables. Mathematical models of the U.S. economy provide information used by the Federal Reserve board to set the prime lending rate. These are just some of the many ways in which mathematics is used in financial applications. In this chapter, we look at the following:

- Simple interest and compound interest
- The exponential function as it relates to compound interest
- Annuities — investments at fixed intervals of time
- Simple stock market simulation
- Trigonometric functions as they relate to analyzing periodic behavior in the stock market

Several applets are introduced that can be used to analyze real-world data sets arising from historical performance of publicly traded stocks. To begin, we analyze one of the oldest financial tools, **simple interest**.

10.1 SIMPLE AND COMPOUND INTEREST

Suppose you put $1,000.00 into a savings account with **fixed interest** (that is, the rate does not vary) of 5%. The amount of interest earned after one year is $50.00 (equal to 5% of $1,000.00). After two years, the amount of interest earned would be $100.00. After 6 months (0.5 of a year), the amount would be equal to \$25.00 (\$1,000.00 × 0.05 × 0.5). From these examples, we can see that the formula for computing **simple interest** is given by

$$I = P \cdot r \cdot t$$

where r is the annual interest rate (expressed as a decimal), t is the time of accumulation (in years), and P is the **principal** (or present value) in the account. The total amount of money in the account, A, is therefore

$$A = P + I = P + P \cdot r \cdot t = P\,(1 + r \cdot t)$$

If you were to take the money out of the account after one year and deposit it into a new account with the same interest, then the principal would be \$1,050.00 and the total interest earned in the second year would be \$52.50 (equal to 0.05 × \$1,050.00). The total in the account would then be \$1,102.50. You gain an additional \$2.50 by reinvesting the amount. Banks make this process more convenient by the process of **compounding**. When interest is compounded, one earns interest on the interest. After two years, the amount is equal to

$$A = 1,050.00 \times (1.05) = 1,000.00 \times (1.05) \times (1.05) = 1000.00 \times (1.05)^2$$

If the interest is compounded each year, for N years, then the amount in the account is equal to

$$A = P\,(1 + r)^N$$

If the time is not exactly an integer number of years, then we can extend the formula to

$$A = P\,(1 + r)^t$$

where t is measured in years.

What happens if we compound the interest more frequently? Suppose the interest is compounded two times a year. After the first six months (0.5 of a year), the amount of interest earned is \$25.00 = \$1,000.00 × 0.05 × 0.5, and the total amount is \$1,025.00. After the second six months you have earned \$25.625 = \$1,025.00 × 0.05 × 0.5, and the total amount is \$1,050.625, which we round down to \$1,050.62. You have gained an additional \$0.62 by compounding twice in one year. The formula for compounding twice a year for t years is given by

$$A = P \cdot \left[\left(1 + \frac{r}{2}\right) \cdot \left(1 + \frac{r}{2}\right) \right]^{t} = P \cdot \left(1 + \frac{r}{2}\right)^{t}$$

If we compound N times a year, we have

$$A = P \left(1 + \frac{r}{N}\right)^{Nt}$$

We can see the dramatic effect of compounding in the applet screenshot that follows.

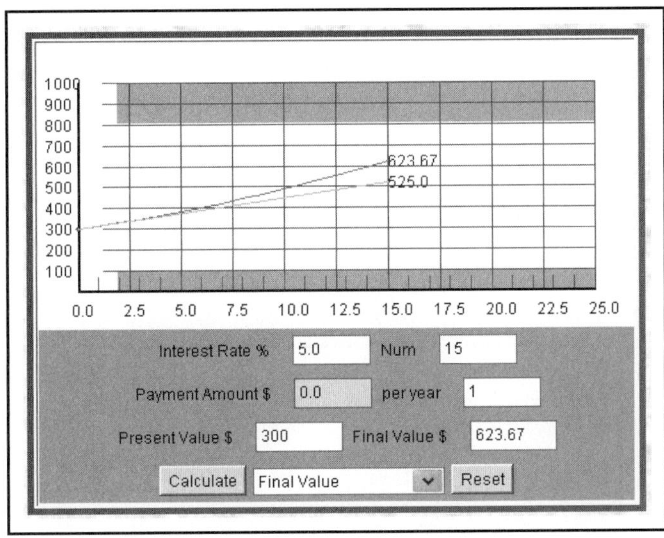

Figure 10.1   Compounding Versus Simple Interest

The top curve in Figure 10.1 gives the amount after compounding $300.00 annually for 15 years, at 5% interest. The lower line gives the amount after 15 years of simple interest at 5%. The difference is $98.67, which is substantial relative to the total amount. Notice that while the simple interest curve is *linear*, the compound interest curve is *non-linear*.

Example 10.1   A savings account has $4,000 with an annual interest rate of 6%. It is kept in the bank for 3 years. How much interest is earned if the account is compounded annually? How much if it is compounded monthly?

Solution

The principal is $P = 4,000$, the interest rate is $r = 0.06$, and the number of years is $t = 3$. When the account is compounded annually, $N = 1$ and the total in the account will be

$$A = 4,000\left(1 + \frac{.06}{1}\right)^{1 \times 3} = 4,764.06$$

So a total of $\$4,764.06 - \$4,000.00 = \$764.06$ is earned in interest. We could also use the Finance Applet to find the total amount (Figure 10.2).

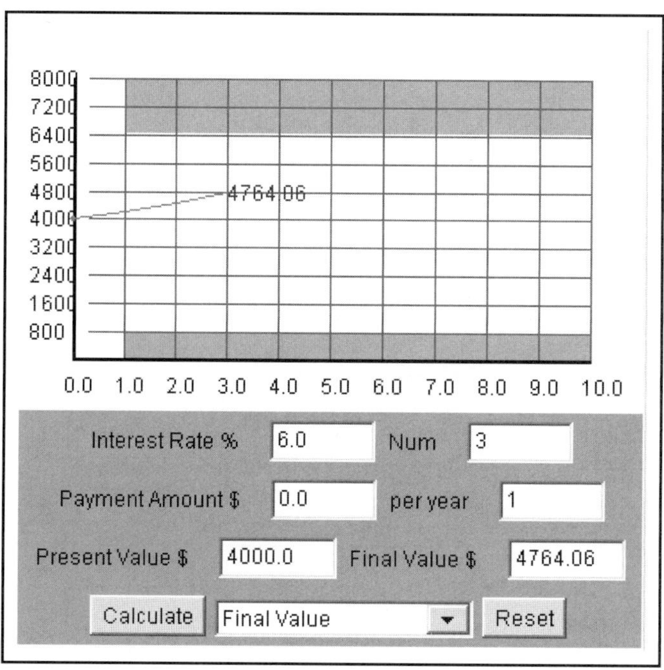

Figure 10.2   Annually Compounded Interest Account

When the account is compounded monthly, $N = 12$ and the total in the account will be

$$A = 4,000\left(1 + \frac{.06}{12}\right)^{12 \times 3} = 4786.72$$

So a total of \$4,786.72 – \$4,000.00 = \$786.72 is earned in interest. We could also use the applet to find the total amount (Figure 10.3).

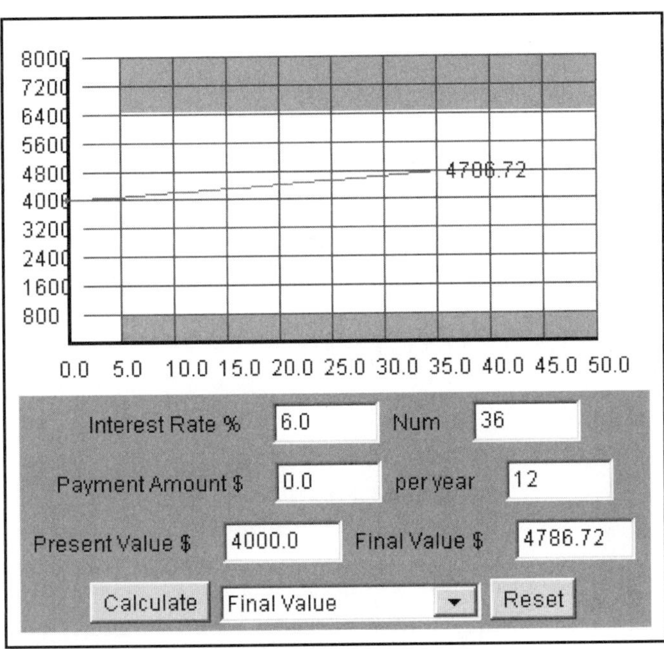

Figure 10.3   Monthly Compounded Interest Account   

Example 10.2   You are offered an interest rate of 5.2% compounded annually or 5% compounded daily. Which is a better deal?

Solution

A simple way to determine which rate is better is to calculate how much is earned with \$100 deposited for one year. For annual compounding you will have

$$A = 100\left(1 + \frac{.052}{1}\right)^{1 \times 1} = 105.20$$

For daily compounding,

$$A = 100\left(1 + \frac{.05}{365}\right)^{365 \times 1} = 105.13$$

So, the annual compounding will be a better deal.   

 **Note:** *You should now complete Activity A in the Financial Applications Module.*

10.2   THE EXPONENTIAL FUNCTION

From the formula for compound interest

$$A(t) = P\left(1 + \frac{r}{N}\right)^{N \times t}$$

we ask what happens as the number of compounding intervals becomes larger and larger. The amount in an account, starting with $P = \$1,000$, $r = 0.05$ and $t = 1$ is shown in Table 10.1 as a function of N, the number of compounding periods.

N	$A(1)$
1	$1,050.00
10	$1,051.14
100	$1,051.26
1,000	$1,051.27
10,000	$1,051.27

Table 10.1   Effect of Increasing the Number of Compounding Periods

Clearly, there is an **effect of diminishing returns**. The difference between $N = 100$ and $N = 10,000$ amounts to $0.01 out of $1,000.00. It is also fairly clear that we are approaching some maximum amount of possible interest for one year. We can also see this by setting $t = 1$ in the formula for compound interest, and expanding the equation as a polynomial.

$$A(1) = P\left(1 + \frac{r}{N}\right)^N$$

With $N = 1$, $A(1) = P \cdot (1 + r)$.

With $N = 2$, $A(1) = P \cdot \left(1 + \frac{r}{2}\right)^2 = P \cdot \left(1 + r + \frac{r^2}{4}\right)$.

With $N = 3$, $A(1) = P \cdot \left(1 + \frac{r}{3}\right)^3 = P \cdot \left(1 + r + \frac{r^2}{3} + \frac{r^3}{27}\right)$.

If we kept on increasing N, we would eventually arrive at the function

$$A(1) = P \cdot \left(1 + r + \frac{r^2}{2} + \frac{r^3}{6} + \frac{r^4}{24} + \cdots\right)$$
$$= P \cdot e^r$$

where $e = 2.718281828459045235602874713527\ldots$

More generally, as a function of t, the formula for **continuous compounding** of interest is

$$A(t) = Pe^{rt}$$

Figure 10.4 gives a comparison of the function e^r and the polynomial approximation with $N = 5$, which is

$$1 + r + \frac{r^2}{2} + \frac{r^3}{6} + \frac{r^4}{24} + \frac{r^5}{120}$$

The solid line is the function $P(r) = e^r$ and the dashed line is the polynomial approximation.

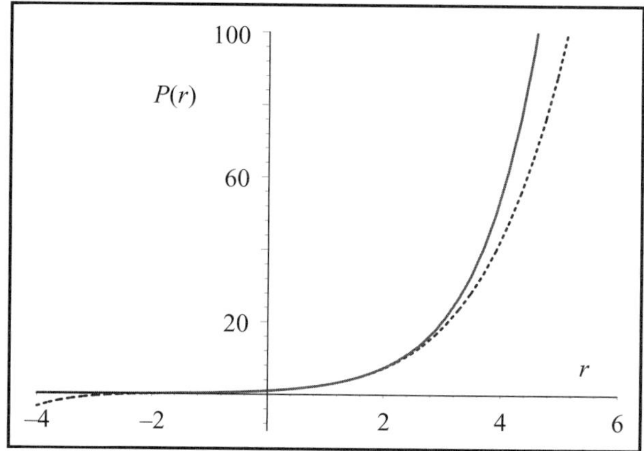

Figure 10.4   Exponential vs. Polynomial Function

One can see that when rt is in the range $[-2, 2]$, the approximation is quite good. For example, if $r = .05$, the 5-term ($N = 5$) approximation is good for $0 \le t \le 80$ years.

Example 10.3   How much interest will we earn if we have \$4,000 deposited for 3 years at 6% compounded continuously? How does that compare with the results of Example 10.1?

Solution

In our formula we will have $P = 4000$, $r = 0.06$, and $t = 3$:

$$A = 4000 \times e^{.06 \times 3} = 4,788.87$$

A total of $4,788.87 – $4000 = $788.87 is earned in interest with continuous compounding. That is $788.87 – $764.06 = $24.81 more than the annual compounding interest and $788.87 – $786.72 = $2.15 more than the monthly compounding.   ❖

 **Note:** *You should now complete Activity B in the Financial Applications Module.*

10.3   ANNUITIES

If one were content to put a sum of money in a savings account, and then leave it for a time, the formulas we have derived would be sufficient. However, what if one decides to make periodic deposits into an account, say, once a month? We would then be contributing to an **annuity.** An annuity is an account to which a series of payments of set size and frequency are added. The formulas we have developed for compound interest are no longer adequate to treat these periodic deposits.

Suppose we have $1,000.00 to put into an account that earns 5% interest compounded monthly. After one month, we decide to put in an additional $100.00. After two months, another $100.00, and so on. After one year, how much is in the account? To answer this, it is easiest to track each investment individually. The initial amount, $1,000.00, after the end of one year, has grown according to the formula

$$1000 \times \left(1 + \frac{.05}{12}\right)^{12}$$

The amount $100.00, deposited at the end of the first month, has grown to

$$100 \times \left(1 + \frac{.05}{12}\right)^{11}$$

The second $100.00, has grown to

$$100 \times \left(1 + \frac{.05}{12}\right)^{10}$$

Adding all of these terms together we get the expression

$$1{,}000 \times \left(1 + \frac{.05}{12}\right)^{12} + 100 \times \left(1 + \frac{.05}{12}\right)^{11} + 100 \times \left(1 + \frac{.05}{12}\right)^{10} + \ldots + 100 \times \left(1 + \frac{.05}{12}\right)^{1} + 100$$

There is a common term, $B = (1 + \frac{0.05}{12})$, which can be factored out, resulting in

$$A = 1{,}000 \times B^{12} + 100 \times (B^{11} + B^{10} + \ldots + B^2 + B + 1)$$

From the identity $(B - 1) \times (B^{11} + B^{10} + \ldots + B^2 + B + 1) = (B^{12} - 1)$, we have

$$(B^{11} + B^{10} + B^{9} + \ldots + B^2 + B + 1) = \frac{(B^{12} - 1)}{(B - 1)}$$

So, the amount A is given by the expression

$$A = 1{,}000.00 \times B^{12} + 100.00 \times \frac{(B^{12} - 1)}{(B - 1)}$$

For arbitrary N and t, we have

$$A(t) = P \cdot \left(1 + \frac{r}{N}\right)^{N \cdot t} + R \cdot \frac{\left(1 + \dfrac{r}{N}\right)^{N \cdot t} - 1}{\left(1 + \dfrac{r}{N}\right) - 1}$$

In this formula, P is the initial deposit, R is the amount deposited at the end of each interest period, N is the number of times the interest is compounded per year, r is the interest rate (in decimal format), and t is the time (in years).

Example 10.4   You open an account to save for holiday gifts. You start with $100 in the account and deposit $30 per month for 11 months. The account pays 4.5% compounded monthly. How much is in the account after 11 months? How much interest have you earned?

Solution

Using the Finance Applet, we enter the following information,

Interest Rate = 4.5   Num = 11

Payment Amount = 30   Per year = 12

Present Value = 100   Final Value = 0

and choose to Calculate the Final Value (Figure 10.5).

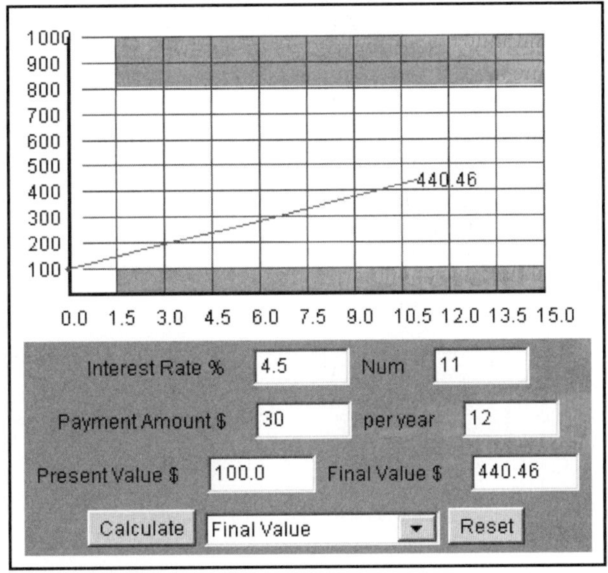

Figure 10.5   Holiday Savings Account

There will be a total of $440.46 in the account. There was $100 + $30 × 11 = $430 deposited, so $440.46 − $430.00 = $10.46 is earned in interest.   ❖

Example 10.5   Suppose a house costs $200,000.00. What are the monthly payments with a 5% down payment, a 30-year **mortgage** (360 monthly payments), and an interest rate of 8.25%? How much interest is paid in all?

Solution

This is very similar to an annuity, except that you are paying off a loan rather than accumulating wealth. After making the down payment, the loan amount is $190,000.00. After one month, if R is the monthly payment, the amount owed is

$$190,000 \times \left(1 + \frac{.0825}{12}\right) - R$$

After two months, the amount owed is

$$\left[190,000 \times \left(1 + \frac{.0825}{12}\right) - R\right] \times \left(1 + \frac{.0825}{12}\right) - R$$

Doing this for 360 months, we must determine the monthly payment, R, which will make the final amount exactly $0.00. This is easily done using the Finance Applet (Figure 10.6). The interest rate, number of payments per year, present value, total, and final value are entered and you choose to Calculate the Payment. The figure also shows the graph of the principal (amount owed) as a function of time.

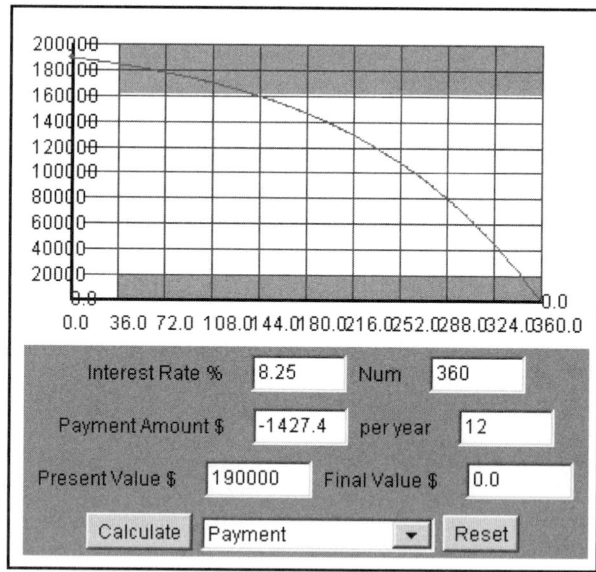

Figure 10.6   Calculating Mortgage Payments with the Finance Applet

In this example, the monthly payment will be $1,427.40. If you take the total amount paid (360 payments of $1,427.40 each), you arrive at $513,864.00, which is $323,864.00 more than the cost of the house! In fact, if you were to put $1,427.40 each month in an interest-bearing account at 8.25% compounded monthly, at the end of 360 months you would have $2,452,560.00, which is more than 10 times the price of the house. This example clearly shows why there is money to be made financing mortgages!   ❖

 **Note:** *You should now complete Activity C in the Financial Applications Module.*

10.4   A SIMPLE STOCK MARKET MODEL

In the previous section on annuities, we made several assumptions. One is that the interest rate is constant (and positive) and another is that we invest a fixed amount of money at regular intervals. Another way in which we might invest regular sums of money is in the stock market, or in mutual funds, where the return is not guaranteed to be positive, and the amount invested may vary depending on stock prices. Furthermore, we may choose to buy, or sell, or do neither, at regular intervals.

A simple stock market applet models this behavior in the following fashion. Initially, the stock price is at a certain amount between $50.00 and $200.00. A horizontal line indicates the price. The price will vary afterward in a manner that is modeled by a fourth order polynomial plus 5% random effects. This means that it is quite difficult to predict prices in the short run, but in the long run, certain trends may appear. Figure 10.7 illustrates this phenomenon.

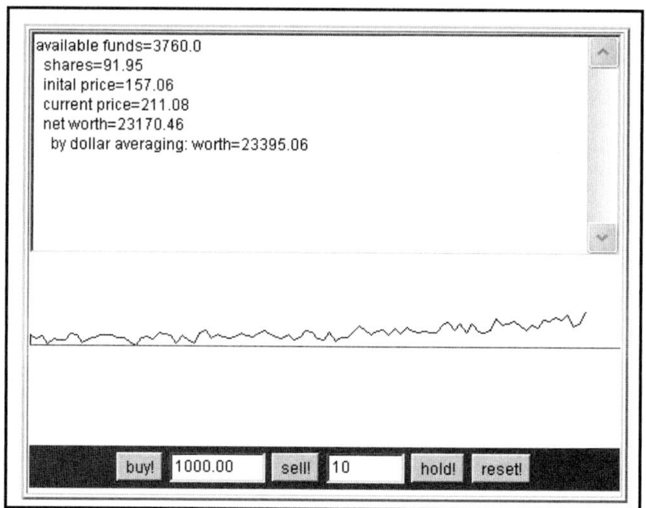

Figure 10.7   Random Price Fluctuations

Initially, one starts with $20,000.00 and buys a certain amount of funds. The default is $1,000.00. Afterward, there are three options:

- Buy, up to the amount of money one has (no loans allowed!).
- Sell, up to the available amount of stock (no options!).
- Hold, that is, do nothing.

Note: *There is a fixed fee, $50.00, for each buy or sell transaction. Furthermore, the time interval between transactions is fixed arbitrarily at 5 business days, that is, on a fixed day of the week.*

One strategy, called **dollar-averaging**, is to buy fixed dollar amounts of mutual funds at regular intervals. If the price is low, more shares are purchased; if the price is high, the total valuation of the holdings is increased. If the overall trend is toward increasing prices, this is a reasonable strategy. The applet measures the value of the portfolio, measures it against the dollar-averaging strategy, and prints out the final profit (or loss), at which time the simulation can be run again. In Figure 10.7, the dollar-averaging amount and the individual's final valuation are the same since the decision was made to buy a fixed amount at each purchase time, that is, the strategies are the same. When the strategies differ, dramatically different results can be seen.

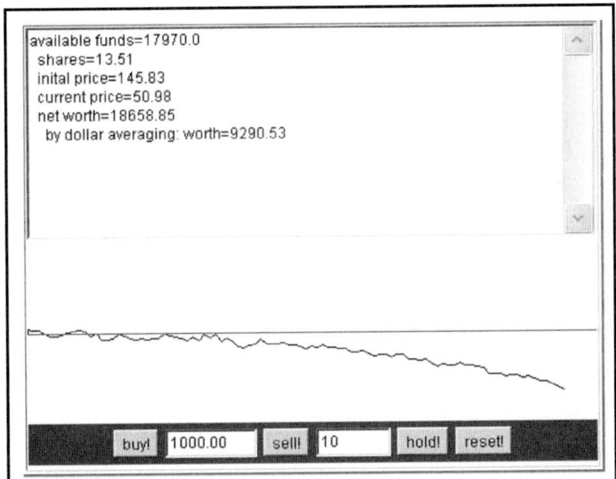

Figure 10.8   The Effect of Different Purchase Strategies

In Figure 10.8, the price has plummeted from an initial price of $119.47 to a final price of $33.16. The decision here was to hold after two initial purchases of $1,000.00, since the price was "obviously" dropping. In a real-world situation, the strategies are much more complex, but in this situation we "saved" over $11,000.00 in losses by basing our buying strategy on short-term trends.

Clearly, one of the most important issues is to analyze the past history of a fund or a stock, and predict future prices to determine our current buying strategies. Analyzing trends in data began, historically, with the development of the least-squares method to analyze astronomical data and compensate for errors in measurement. In the next section, we look at a non-linear least-squares method and its role in analyzing trends in data.

10.5   TREND ANALYSIS

The actual performance of a real stock can be very complex (Figure 10.9).

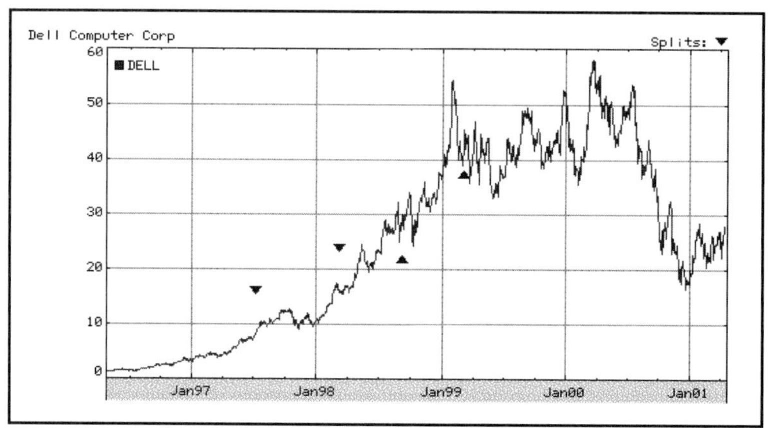

Figure 10.9   Five-Year Price History for Dell Stock

How on earth are we supposed to model this kind of behavior? In the method of least squares we found the "best fit" for a straight line of the form $y = mx + b$ but in non-linear least-squares we have more functions available to use to fit our data. Instead of $\{1, x\}$ we might have $\{1, x, x^2\}$ or $\{1, x, x^2, x^3\}$. The set of functions we use to model the behavior is called a basis, and we are free to choose whatever functions we wish. For example, if we were to use the functions $\{1, x, x^2, x^3, \sin(x), \cos(x), e^x\}$, then we would try to find the best fit for a function of the form

$$y = a_1 \cdot 1 + a_2 x + a_3 x^2 + a_4 x^3 + a_5 \sin(x) + a_6 \cos(x) + a_7 e^x$$

The choice of the basis functions is somewhat of an art, but simply stated, the goal is to get the best fit with the fewest possible functions. In Figure 10.10, the Non-Linear Least-Squares Applet is shown with data from Figure 10.9, and fitting to the same basis functions listed above.

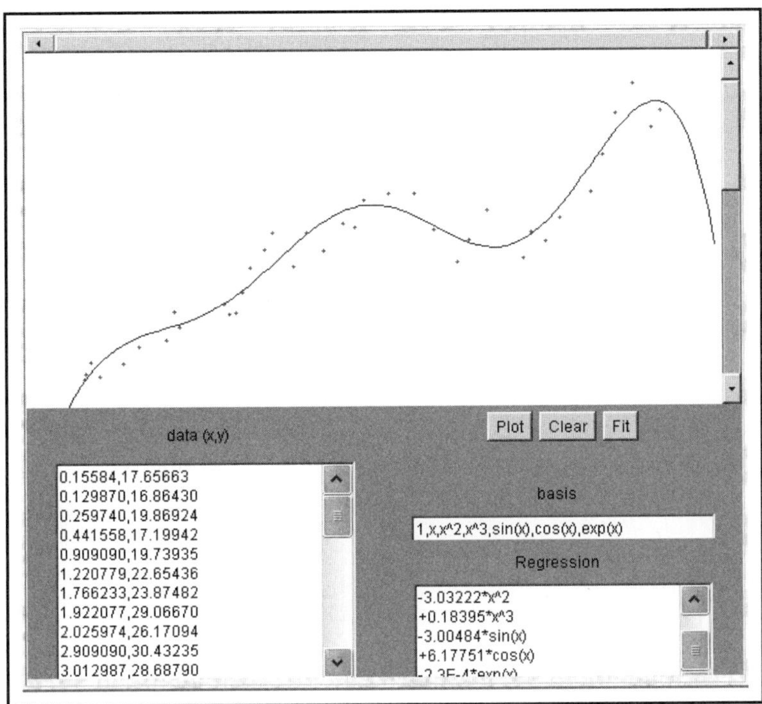

Figure 10.10   Non-Linear Least-Squares Applet

One can see that the general behavior is picked up, but that a huge drop is predicted at the end of the interval. Without a rational reason for picking the particular basis functions used, and a rigorous justification of the values of the coefficients obtained, it would be risky to base future purchases on the results of this applet!

Note: *You should now complete Activities D and E in the Financial Applications Module.*

FURTHER EXPLORATIONS

For further explorations in the area of financial applications please visit the URL

http://www.finitemathtutor.com/explore/chapter10/

EXERCISES

Exercises 10.1–10.3:

A bank account starts with $400, pays 6.2% annual interest, and remains in the bank for 5 years.

Exercise 10.1   How much interest is earned if the account is compounded annually?

Exercise 10.2   How much interest is earned if the account is compounded daily?

Exercise 10.3   How much interest is earned if the account is compounded continuously?

Exercise 10.4   You are given a choice of a savings account that pays 3.8% compounded daily or one that pays 3.9% compounded annually. Which has a higher yield?

Exercises 10.5 and 10.6:

A car is financed with a loan for $21,000 for 3 years at 5.9% annual interest rate with monthly payments.

Exercise 10.5   How large are the monthly payments?

Exercise 10.6   How much interest is paid in all?

Exercises 10.7 and 10.8:

You open a savings account to save for a pair of diamond earrings. You start with $50 in the account and you deposit $10 every week for two years. The account pays 6.6% compounded weekly.

Exercise 10.7   How much is in the account in two years?

Exercise 10.8   How much interest did you earn?

Exercises 10.9 and 10.10:

You have a college savings account to which you make annual contributions of $2,500. The account pays 8% compounded annually.

Exercise 10.9   How much is in the account after 17 years?

Exercise 10.10   How much interest did you earn?

SAMPLE QUIZ

Questions 10.1–10.3:

An account starts with $8,000, pays 4.5% annual interest, and remains in the bank for 4 years.

Question 10.1   How much interest is earned if the account is compounded annually?

Question 10.2   How much interest is earned if the account is compounded daily?

Question 10.3   How much interest is earned if the account is compounded continuously?

Question 10.4   You are given a choice of a savings account that pays 4.3% compounded monthly or one that pays 4.25% compounded annually. Which has a higher yield?

Questions 10.5 and 10.6:

A car loan for $28,000 is financed for 5 years at 7.2% annual interest rate with monthly payments.

Question 10.5   How large are the monthly payments?

Question 10.6   How much interest is paid in all?

Questions 10.7 and 10.8:

You open a savings account to save for a trip to Hawaii. You start with $500 in the account and you deposit $100 every month for four years. The account pays 7% compounded monthly.

Question 10.7   How much is in the account after four years?

Question 10.8   How much interest did you earn?

Questions 10.9 and 10.10:

You have a retirement fund to which you make annual contributions of $5,000. The account pays 9% compounded annually.

Question 10.9   How much is in the account after 20 years?

Question 10.10   How much interest did you earn?

Applets

Slope Applet

Goal: To see that any two points are enough to determine the unique slope of a straight line.

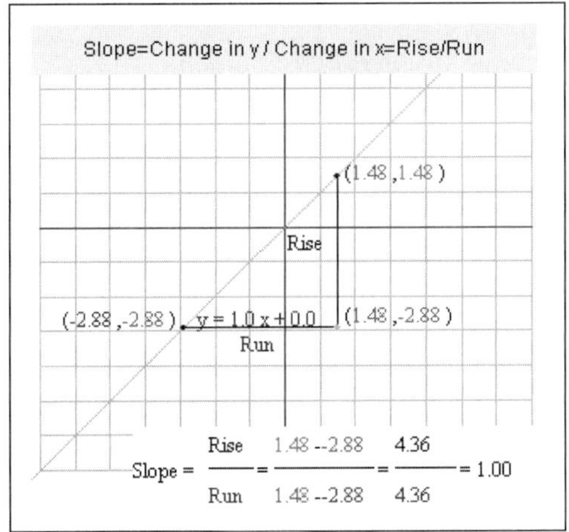

Figure A.1   Graphing Applet

Usage:

- Enter a slope and *y*-intercept (if dialog box is present).
- Drag the corner of the triangle farthest away from the line to see how the slope recalculates to the same value.

Line Graphing Applet

Goal: To graph one or two lines given either the slope and *y*-intercept or two points on the line. The appearance of this applet will vary depending on the input required.

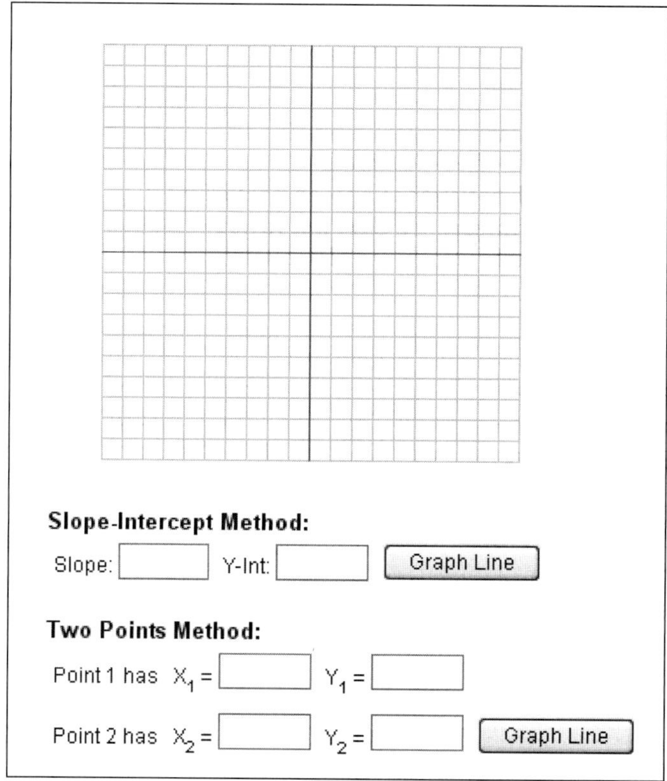

Figure A.2   Line Graphing Applet

Usage:

- If you are given the slope and the intercept, enter the values into the boxes and click the Graph Line button.
- If you are given two points on the line, enter those values into the boxes and click the Graph Line button.

nteractive Least-Squares Applet

Goal: To visualize the regression line, as it depends on data, in real time. This is a dramatic example of the mathematical concept of continuous dependence.

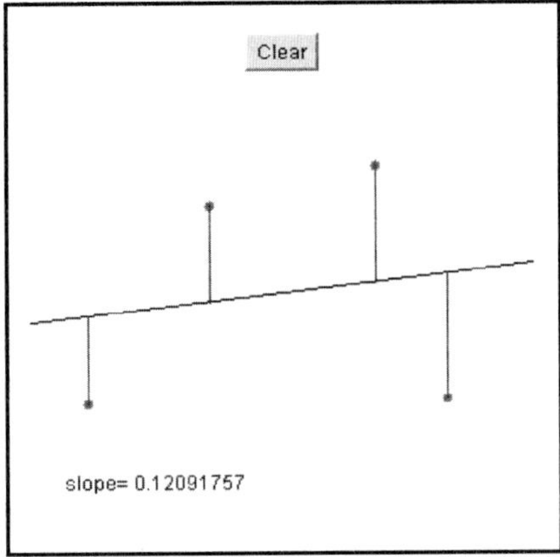

Figure A.3   Interactive Least-Squares Applet

Usage:

- Click two or more times at different places in the applet. Circles indicate the location of the data points, the thicker black line represents the regression line, and the thin black lines represent the error.
- If you click on a data point, you can drag it to a new location. The regression line and the slope are automatically updated.
- The slope of the line is indicated in the lower left corner.

Reaction Time Applet

Goal: To construct individualized data sets which can then be analyzed with the Least-Squares Applet.

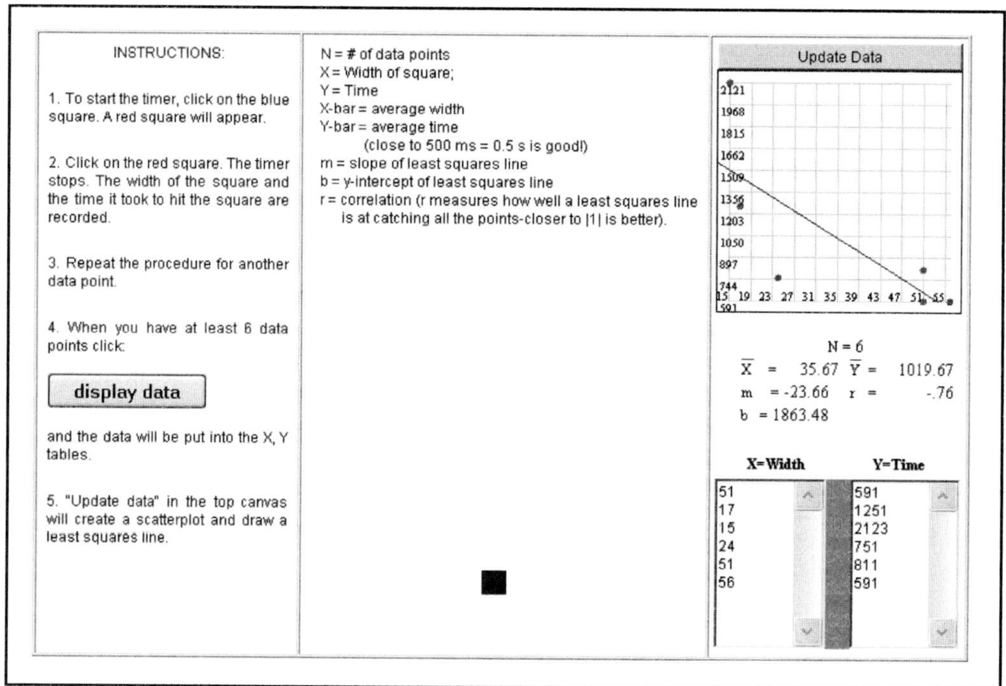

Figure A.4   Reaction Time Applet

Usage:

The instructions for this applet are contained within the applet itself.

Non-Linear Least-Squares Applet

Goal: To see how the choice of basis functions affects the regression curve.

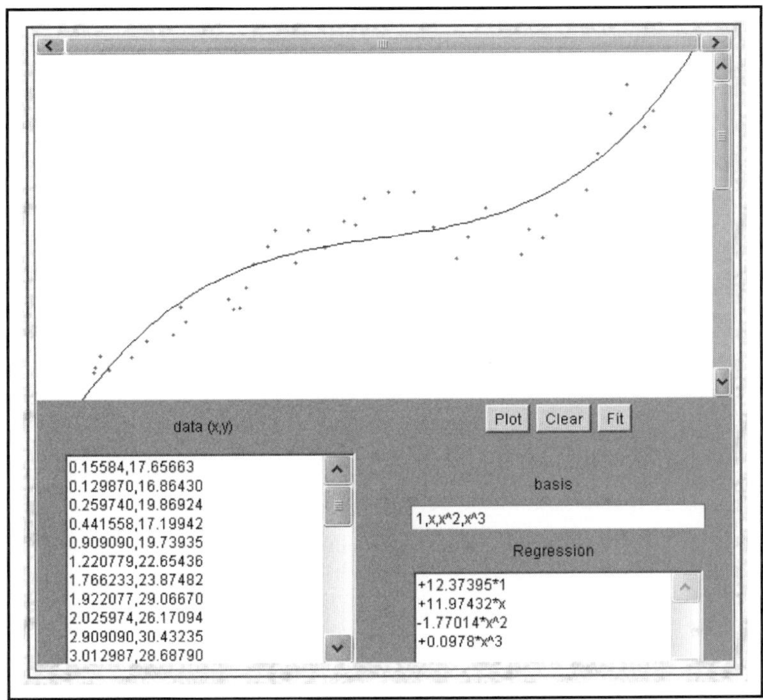

Figure A.5   Non-Linear Least-Squares Applet

Usage:

- Enter the data as pairs of comma-separated x and y points in the text area in the lower left part of the applet. Press the "Plot" button to see the data points.
- Enter the choice of basis functions, separated by commas.
- Press the "Fit" button to find the regression curve for the chosen basis functions.
- The coefficients of the basis functions are shown in the box in the lower right corner.

Gauss-Jordan Applet

Goal: To visualize the solution of a linear system of equations by executing elementary row operations.

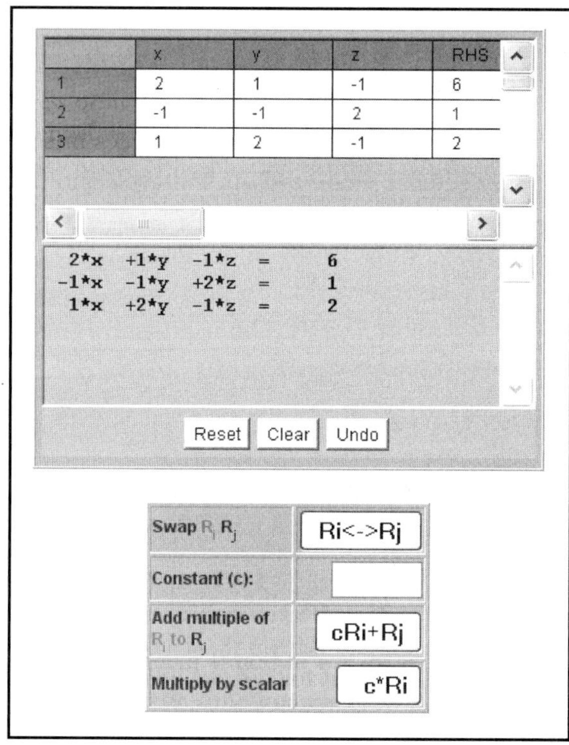

Figure A.6   Gauss-Jordan Applet

Usage:

- To change the contents of a cell, click on the cell and type a numerical value. The system of equations below the table will automatically update.
- To change variable names, click on the column and enter the variable name.
- To change the right-hand side of the equation, click on the far right column and enter the values.

- To select a row for the purposes of elementary row operations, click on the far left column. The first row, R_i, is highlighted in blue and the second row, R_j, is highlighted in red.
- To interchange two rows, click the first button in the table.
- To add a multiple of one row to another, select two rows and enter the constant into the text field (second row of the table) and click the button below the box in the table.
- To multiply a row by a constant, select the row to be multiplied. Enter the constant. Click the fourth button in the table.
- To deselect a row, click it.

Method of Corners Applet

Goal: To visualize the principle that the maximum and minimum of a linear function, subject to linear constraints, occur on the boundary of the feasible region. By putting in non-linear objective functions or non-linear constraints, one can see how important the assumption of linearity is.

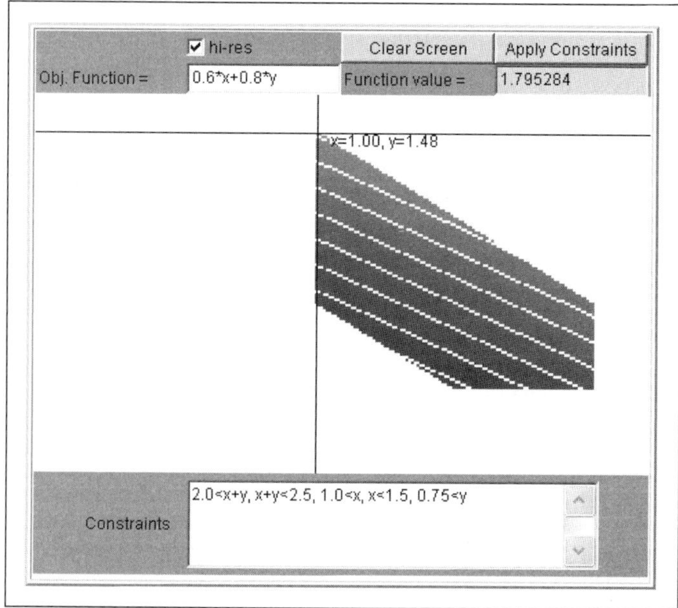

Figure A.7   Method of Corners Applet

Usage:

- Enter the objective function as a function of x and y.
- Enter the constraints, separated by commas.
- Implied multiplication is not allowed. $2x$ must be entered as $2*x$ or the function will not graph.
- Click the "Apply Constraints" button.
- Check the "Hi-Res" box for better quality graphics.

- Click on any point in the interior to see the value of the objective function at the cross-hairs. As the objective function increases in value, the shading on the applet lightens.
- Curves where the objective function is constant are indicated by white stripes.
- Click the "Clear Screen" button to begin again.

Simplex Applet

Goal: To see how the simplex algorithm works in three dimensions by appropriately choosing the right corners of the feasible set. This applet can be used to solve linear programming problems in three variables with an arbitrary number of constraints.

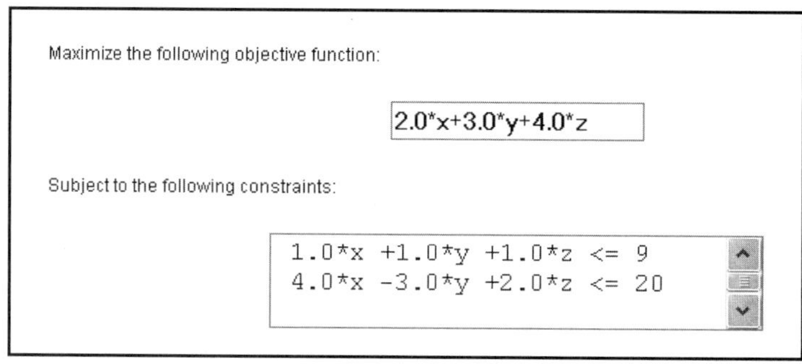

Figure A.8a   Simplex Applet — Entering the problem

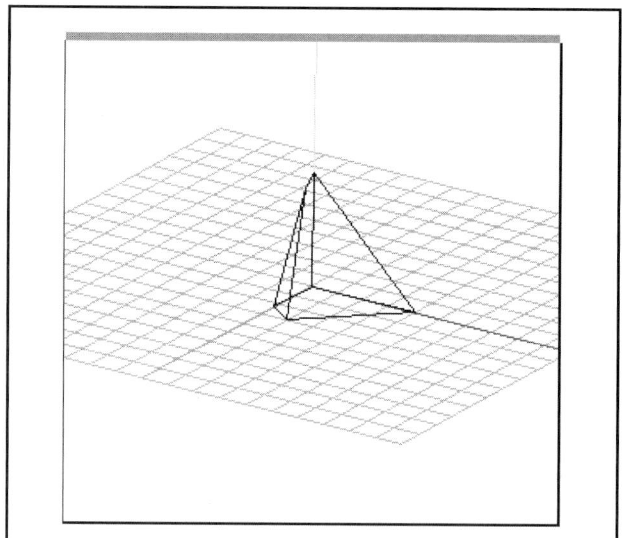

Figure A.8b   Simplex Applet —Viewing the Feasible Region

Build the Simplex Tableau

x	y	z	u0	u1	P	C
1.0	1.0	1.0	1.0	0.0	0.0	9.0
4.0	-3.0	2.0	0.0	1.0	0.0	20.0
-2.0	-3.0	-4.0	0.0	0.0	1.0	0.0

Figure A.8c   Simplex Applet — the Simplex Tableau

Usage:

- Enter the Objective Function and the Constraints for the Simplex Algorithm.
- View the Feasible Set by clicking the button. Clicking anywhere in the graph's box and dragging the mouse will allow view of the region to be rotated.
- Generate the initial simplex tableau by clicking the button.
- Begin the Simplex Algorithm by clicking the mouse under the pivot column, then selecting the pivot row by examining the ratios.
- Click the "Show Vertex" button (if present) to visualize the sequence of corners on the feasible region graph.

Set Building Applet

Goal: To visualize the actions of union and intersection on a set which the student has created individually.

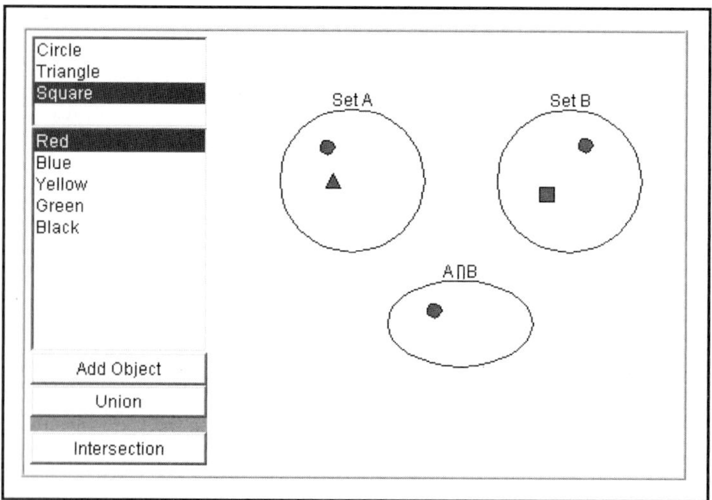

Figure A.9   Set Union and Intersection Applet

Usage:

- In the Set Building Applet, one can choose 3 geometric shapes and 5 colors for a total of 15 possible unique objects.
- To add an object in the Set Builder Applet, press the "Add Object" button.
- Drag the newly created object to one of the circles.
- To see the union and intersection of the sets, press "Union" and "Intersection" respectively.

Venn Diagram Applet

Goal: To interactively create representations of unions, intersections, and complements.

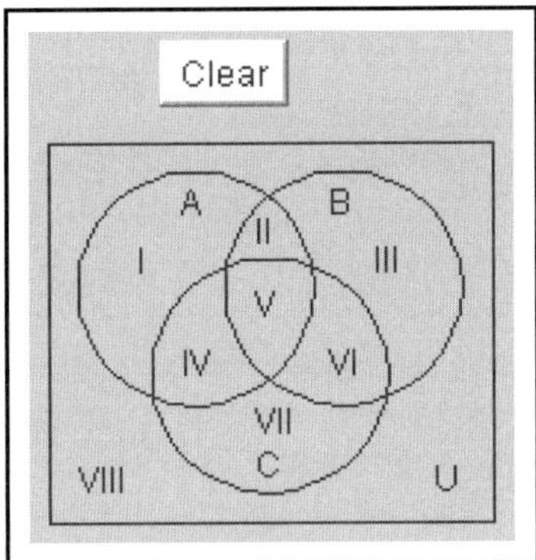

Figure A.10   Venn Diagram Applet

Usage:

- Click on each region to highlight it. When finished, press "Submit" button (if present).
- To remove highlighting from a region, click on the highlighted region again.
- To reset the applet, press the "Clear" button.

Urn Applet

Goal: To visualize the relationship between random selection, relative frequency, conditional probability, and theoretical probability. One can see how slowly the relative frequencies converge to the theoretical probabilities.

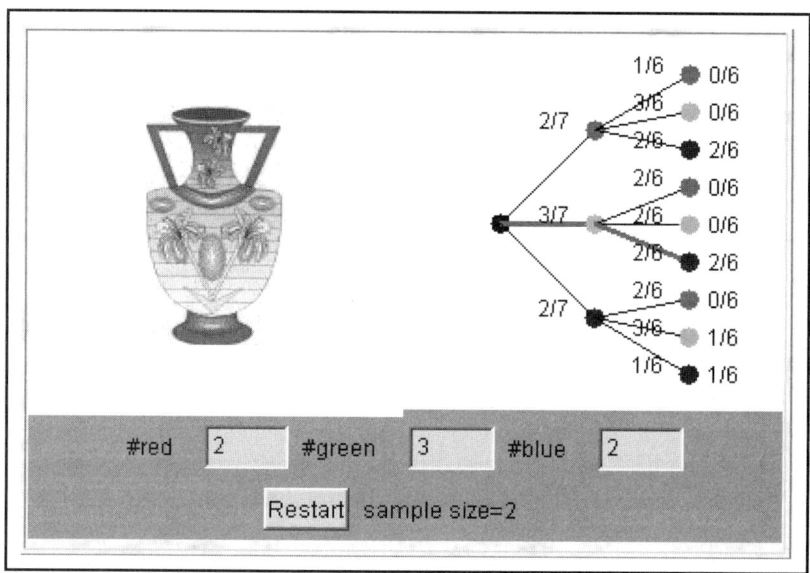

Figure A.11   Urn Applet

Usage:

- Given the numbers of red, green, and blue balls in the urn, the relative probabilities of choosing a red, green, or blue ball are displayed in the tree diagram at the right of the applet.
- Click on the urn, and the first ball is displayed.
- Click again on the urn, and the second ball is displayed. The branch of the tree diagram is highlighted in red, and the relative frequencies are updated.
- To reset, press "Restart."

Statistics Applet

Goal: To interactively compute and display the various statistics for a data set which the student can create. This applet allows one to first estimate the statistical quantities before showing their true values.

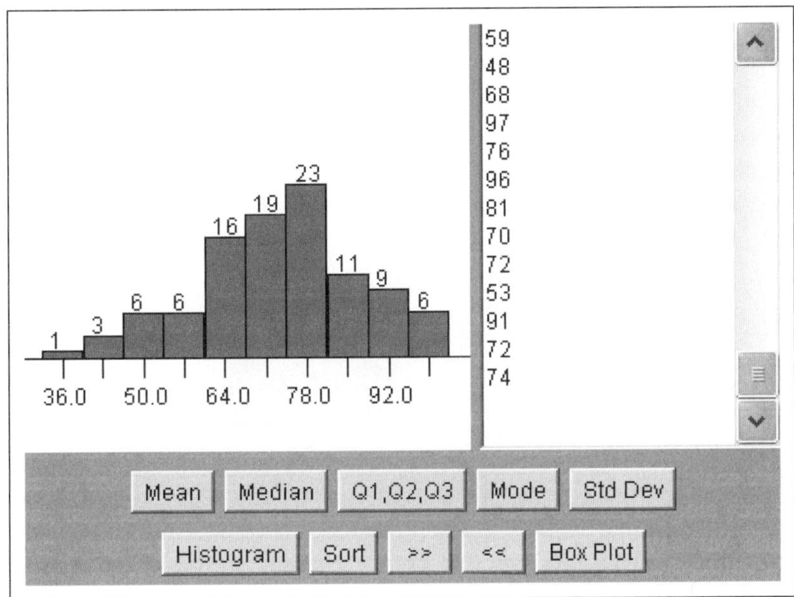

Figure A.12   Statistics Applet

Usage:

- Enter the data into the text area at the right of the applet.
- Press the "Histogram" button to view the histogram. Press ">>" to increase the number of rectangles or press "<<" to decrease the number of rectangles.
- The "Mean," "Median," "Q1,Q2,Q3," "Mode," and "Std Dev" buttons will display the appropriate quantity on the histogram for the data.
- Press "Sort" to sort the data (in ascending or descending order).
- Press "Box Plot" to display a box plot for the data.

Data Generator Applet

Goal: To link the concepts of histogram, population distribution, and sample distribution in an interactive way. This applet is an incredibly powerful tool for creating random data with prescribed mean and standard deviation.

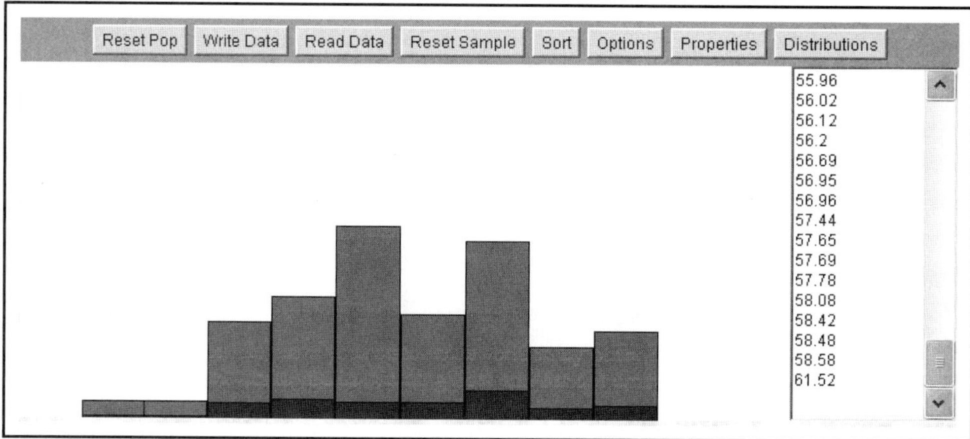

Figure A.13a   Data Generator Applet — Main Window

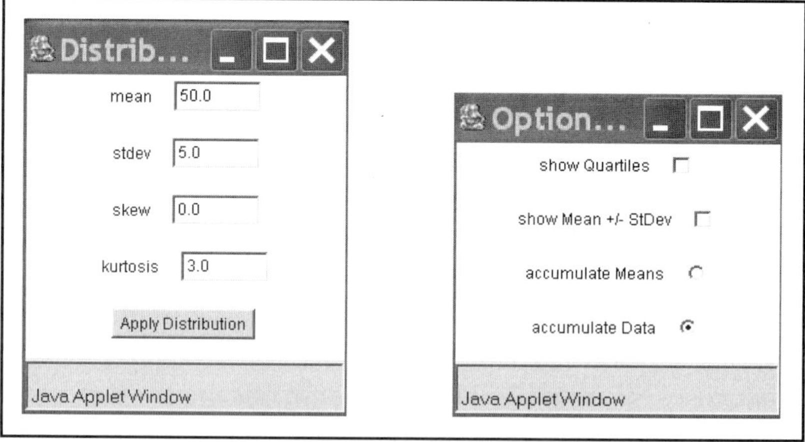

Figure A.13b   Data Generator Applet Dialog Boxes

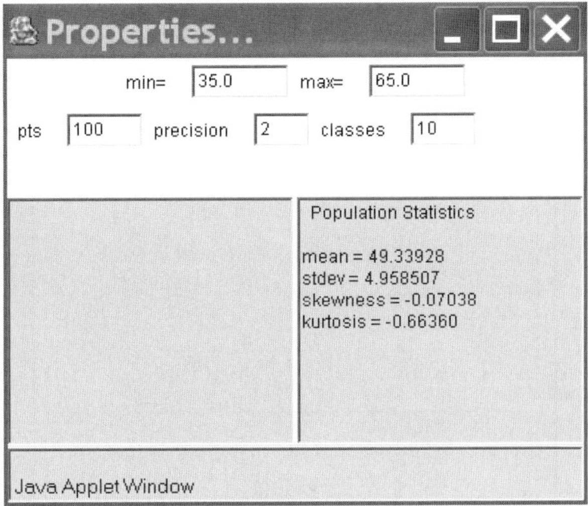

Figure A.13c   Data Generator Applet Dialog Box

Usage:

- In the "Properties" dialog box, set the minimum and the maximum of the data points, set the number of data points and the precision, and set the number of classes (rectangles) for the histogram.
- Press the "Reset Pop" button to reset the population to a uniform normal distribution.
- Use your mouse to drag the top of each column to the desired height, creating a histogram of the population, shown in red.
- Press the "Write Data" button to create a random sample, which is recorded in the text area at the right side of the applet. The histogram of the sample is shown in blue.
- You can "sort" the data in ascending or descending order by pressing the "Sort" button.
- Notice the "X" button in the upper right corner does not close the dialog boxes!
- You can reset the sample population by pressing the "Reset Sample" button.

Correlation-Outliers Applet

Goal: To demonstrate the concept of an outlier in two dimensions.

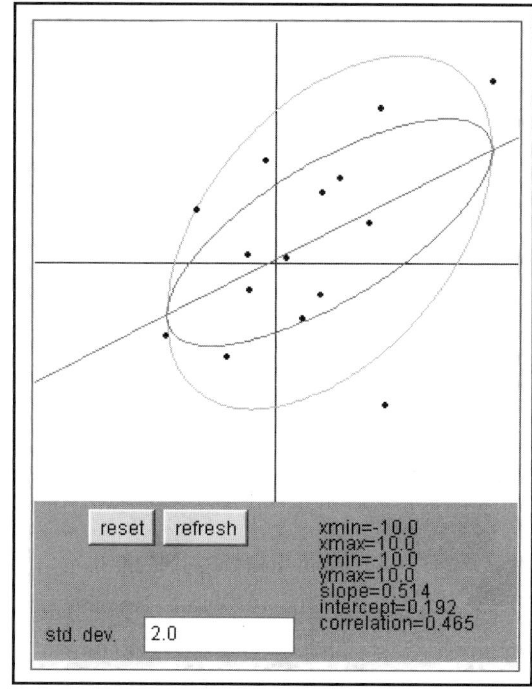

Figure A.14   Correlation-Outliers Applet

Usage:

- Enter the number of standard deviations for the outliers into the text box at the bottom of the applet.
- The red straight line is the regression line, $y = mx + b$.
- If you subtract the regression line values from the data values, and calculate the standard deviation, σ, of the differences $y_i - (mx + b)$, then the red ellipse has a major axis equal to the range of the data $x_{max} - x_{min}$ and a minor axis equal to σ.
- The green ellipse has a minor axis equal to a multiple of σ.

Normal Distribution Applet

Goal: To visualize the relationship between the normal and the standard normal curves. To visualize the relationship between intervals, standard deviations, and areas.

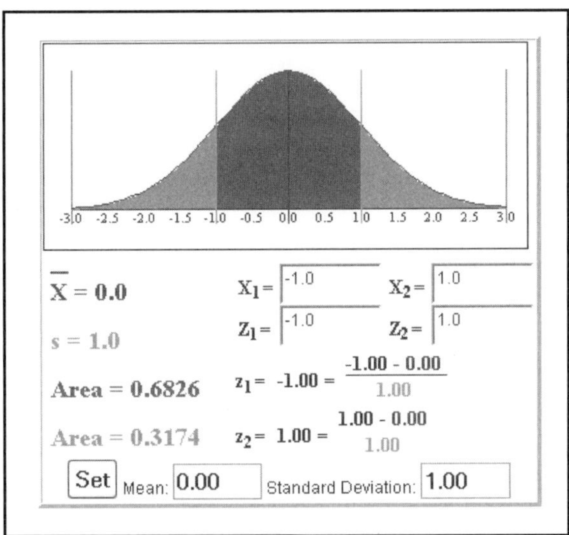

Figure A.15   Normal Distribution Applet

Usage:

- Enter the mean and the standard deviation in the text fields at the bottom of the applet. Press the "Set" button.
- Enter the X_1 and X_2 values to set the lower and upper limits of the interval. Press Enter on your keyboard after entering each of these values, rather than using the Tab key on your keyboard. The equivalent Z-values are displayed.
- You can drag the vertical green lines to set the limits of the interval also.
- You can enter standardized Z-values and see the equivalent X-values.
- The area inside the interval is shaded in green and the area outside the interval is shaded in red.

Coin Toss Applet

Goal: To demonstrate the approach of the binomial distribution to the normal distribution.

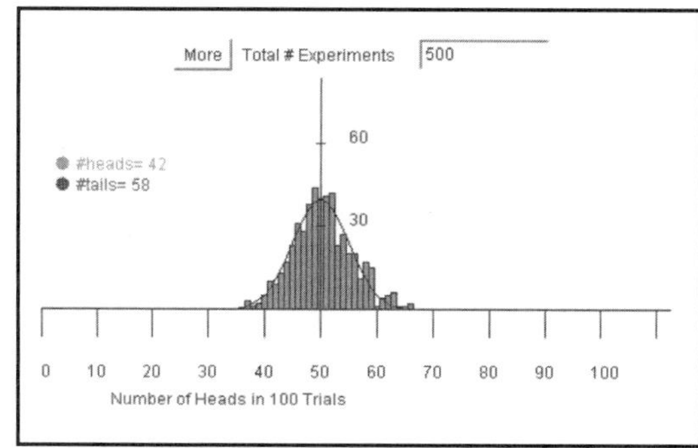

Figure A.16   Coin-Toss Applet

Usage:

- Click the "More" button to increase the total number of experiments. The number of experiments after each button press is given in the text area at the top of the applet.
- The number of heads in each experiment is given in red, the number of tails in blue.
- The last line of text gives the number of trials in each experiment.
- By clicking the mouse above a vertical bar, the number of heads in that category is listed.

The smooth curve that overlays the histogram is the normal distribution with the same mean $\mu = Np$, and standard deviation $\sigma = \sqrt{Npq}$ as in the binomial coin-toss experiment, where N is the number of trials and the probability of success is given by $p = q = \frac{1}{2}$.

Finance Applet

Goal: Given all of the information, except for one parameter, one can find the value of the remaining parameter. This is modeled after the TVM (total value of money) function on graphing calculators.

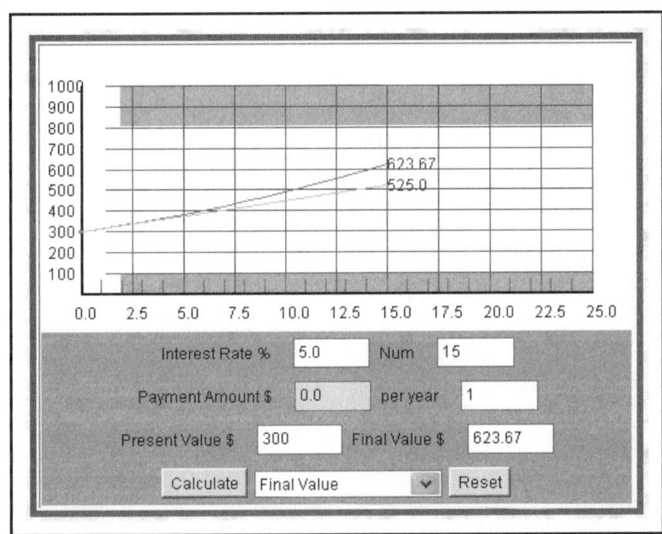

Figure A.17   Finance Applet

Usage:

- Enter the five of the six quantities that are known: interest rate, the total number of payments, the payment amount, the interest periods per year, the principal, and the final value.
- Select one of the following:

    Final Value, Interest, Number of Payments, Present Value or Payment

- Press the "Calculate" button.

Stock Market Applet

Goal: To illustrate the complexity of even simple financial models, and the difficulty of devising strategies based on observing "historical data" over a limited time period. This is why there are not very many successful day-traders.

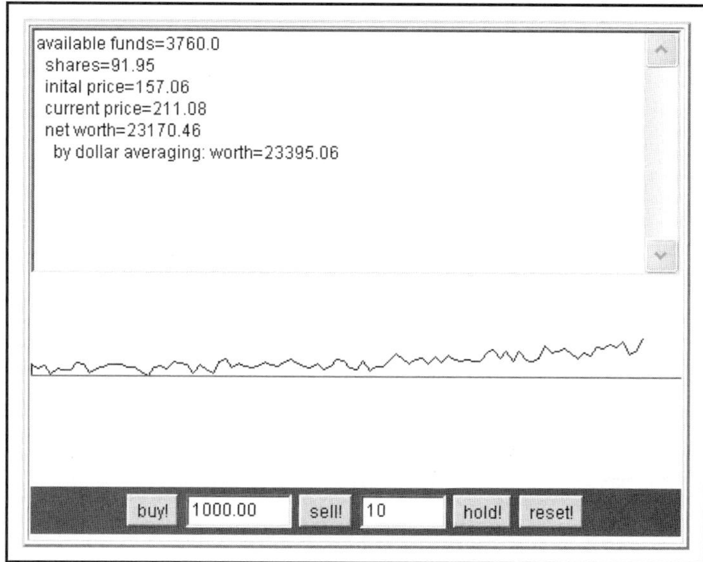

Figure A.18   Stock Market Applet

Usage:

- When the simulation begins, you have $20,000.00 in your account. You can choose to "buy," "sell," or "hold."
- Each transaction costs $50.00. (Holding does not cost anything.)
- You can sell only as much stock as you have, and you can buy only as much as you have funds for.
- At the end of the simulation, the stocks are sold, and the net amount of funds is computed and compared to a "dollar-averaging" strategy.

Image Digitizer Applet

Goal: To enable one to digitize images from the Web in order to extract numerical data from images and graphs.

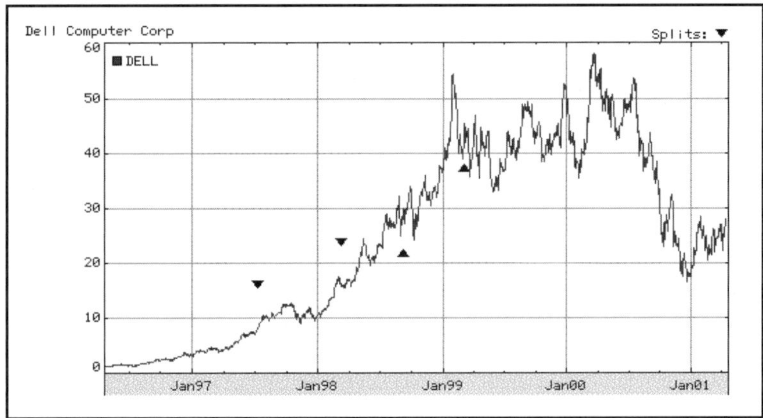

Figure A.19   Image Digitizer Applet

Usage:

- Enter the URL address of an image.
- Enter the x and y coordinates of a point in the lower left part of the graph into the two boxes next to the "Set Lower Left" button. Here we have x in units of time and y in units of price.
- Click on the image in the approximate location of the point entered in the boxes with the mouse. Press the "Set Lower Left" button.
- Use a point in the upper right corner in a similar fashion to set the upper right values.
- Press the "Clear" button to remove any data.
- Click on a sequence of locations on the curve, as many as you want. The data appears in the text area in the lower left. You can copy and paste the data into other applications.

Glossary

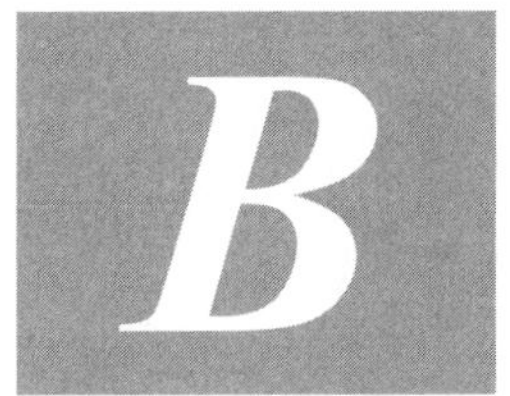

- **Absolute maximum** – a function value that is greater than or equal to all other function values in the range.

- **Absolute minimum** – a function value that is less than or equal to all other function values in the range.

- **Annuity** – a series of payments of set size and frequency (usually at the beginning or end of an interest period).

- **Augmented matrix** – a matrix formed by adding an additional column, corresponding to the right-hand side of a linear equation, to the coefficient matrix.

- **Average** – the sum of a set of real numbers, divided by the total number of values.

- **Binomial distribution** – a probability distribution arising from binomial experiments which have the following properties:
  - A fixed number of trials, N
  - A fixed probability of success (the outcomes are independent)
  - Only two possible outcomes, success and failure

- **Boolean logic** – a system of logic, used in all modern computers (where $0 = F$ (false) and $1 = T$ (true)), and based on the three operations of
  - *And*: (T and T) = T, (T and F) = F, (F and T) = F, (F and F) = F
  - *Or*: (T or F) = T, (F or T) = T, (F or F) = F, (T or T) = T
  - *Not* (complement): (not T) = F, (not F) = T

- **Calculus of variations** – the study of minimizing or maximizing integral functions of several variables.

- **Cardinality** – the number of elements in a set.

- **Categories** – subsets, which are determined by inequalities, ranges, or intervals.

- **Central limit theorem** – the result that a distribution of sample means approaches a normal distribution as the number of samples becomes arbitrarily large.

- **Closed** – describes a set that contains all its possible limits.

- **Coefficient matrix** – the matrix that results when a system of equations is reduced to matrix form. The ith row and j^{th} column of the coefficient matrix is equal to the coefficient multiplying the jth variable in the ith equation.

- **Collection of objects** – an unordered set of objects.

- **Column matrix** – a matrix consisting of exactly one column.

- **Combination** – the number of ways an event can occur, without taking order into account.

- **Complement** – the complement of a set consists of all elements which are *not* members of that set.

- **Compound interest** – a process in which interest is added to the principal at the end of each compounding period.

- **Compounding** – adding the interest to the principal prior to calculating the interest over the next period.

- **Conditional probabilities** – if two events are not independent, then their probabilities are related. $P(A|B)$ is the probability of A occurring, given that B has already occurred. Similarly, $P(B|A)$ is the probability of B occurring, given that A has already occurred.

- **Continuous compounding** – this is the case of compounding in the limit as the interest period approaches zero.

- **Continuous distribution** – a probability distribution whose events have probabilities with values on an entire range of probabilities.

- **Convex** – the shape of a figure containing all points on any line segment connecting points on its boundary.

- **Correlation coefficient** – a measure of the independence of two random variables. A value of $+1$ or -1 indicates perfect linear correlation. A value of 0 indicates the absence of any correlation.

- **Decomposition** – expression of a linear system of equations in matrix form.

- **Degrees of freedom** – the number of unrestricted, independent, random variables present in a problem. Degrees of freedom also refers to the number of undetermined coefficients present in a function.

- **Demand curve** – the relationship between the amount of a commodity consumers are willing to purchase and the price.

- **Dependent variable** – a variable whose value must be computed through a function.

- **Discrete distribution** – a probability distribution with a finite number of outcomes.

- **Dollar-averaging** – purchasing a fixed dollar amount of stocks or bonds at regular intervals in order to minimize the effects of market variability.

- **Domain** – the set of values for which a function is defined.

- **Effect of diminishing returns** – when an effect becomes less noticeable even though it is applied more often.

- **Empirical probability** – the relative frequency that an event occurs, calculated by actual observation.

- **Empty set** – the set containing no elements (analogous to the additive identity).

- **Equations** – a statement (in the form of an equality) between values on the left side of the equality and values on the right side of the equality.

- **Error** – the difference between predicted and observed measurements. Also, random (and non-reproducible) differences in measurements occurring from the measurement process itself.

- **Event** – a subset of the sample space.

- **Exclusive or** – logical operation describing elements which are in one set or another, but *not both*.

- **Expected value** – the average value of a random variable, weighted by its probability of occurrence.

- **Experiment** – any activity that has an observable result.

- **Factorial** – the product of all integers in sequence, from 1 to N, is N-factorial, written $N!$

- **Feasible region** – a set of values in the domain of the objective function that satisfy all the inequalities (constraints).

- **Finite** – a set whose elements can be listed by a process that terminates.

- **Finite number of outcomes** – an experiment for which all possible outcomes can be listed individually.

- **Fixed interest** – the case where interest rates do not vary.

- **Function** – a relationship (or rule) that ascribes one and only one value in the second set (range) to each member of the first set (domain).

- **Gauss-Jordan elimination** – the process of using elementary row operations to find the solution of a system of linear equations.

- **Graph** (or plot) – a diagram that illustrates the relationship between two sets of numbers, one plotted along the horizontal axis (abscissa) and one along the vertical axis (ordinate).

- **Grouped frequency distribution** – frequency of occurrence of values within categories.

- **Horizontal line** – a line with zero slope. A set of all (x, y) points with the same ordinate (y-values).

- **Identity matrix** – a matrix consisting of ones along the main diagonal, and zeroes everywhere else.

- **Inclusive or** – the set operation describing elements that are in one set or another, possibly both.

- **Independent events** – if the probability of one event occurring is unchanged by the occurrence of another event, then these events are independent.

- **Independent variable** – a variable whose value can be set independently of other variables in a functional relationship.

- **Inequality constraints** – a constraint which is of a particular form, that of an inequality.

- **Infinite number of outcomes** – an experiment for which has an infinite number of possible outcomes.

- **Intercept** – the coordinate where a curve intersects a coordinate axis.

- **Interpretation of data** – deducing relationships between sets of data by looking at individual and joint statistics.

- **Interquartile range** – the numerical difference between the third and first quartiles.

- **Intersection** – the set of all elements belonging to one set *and* another.

- **Inverses** - $\mathbf{B}$ is the inverse of a matrix $\mathbf{A}$ if $\mathbf{BA} = \mathbf{AB} = \mathbf{I}$ where $\mathbf{I}$ is the identity. Only square matrices have inverses.

- **Law of large numbers** – the process of approaching a normal distribution as the number of trials becomes arbitrarily large.

- **Least-squares method** – the method of determining the relationship between two sets of data in such a way that the sum of the squares of the error, measured by the difference between predicted and observed values, is minimized.

- **Limit** – the process in which one quantity approaches arbitrarily close to another quantity.

- **Linear depreciation** – expressing the loss in value of a quantity by means of a linear relationship.

- **Linear function** – a relationship between variables whose graph is a straight line.

- **Linear programming problem** – A problem in which one wishes to maximize or minimize a linear function subject to a set of linear inequalities.

- **Linear regression** – the process of finding a regression line for a given set of data

- **List** – an ordered collection of objects.

- **Market equilibrium** – a condition where supply of a commodity equals demand.

- **Mathematical modeling** – expressing the relationships between quantities (or variables) through mathematics (equations or inequalities).

- **Matrix** – a rectangular array of values.

- **Matrix addition** – the sum of two matrices, taken element by element.

- **Matrix equality** – two matrices are equal if and only if they are the same size, and they have the same elements at each row and column.

- **Matrix inverse** – The matrix inverse, $\mathbf{A}^{-1}$, is a matrix such that when multiplied (on the left or the right) by the original matrix, $\mathbf{A}$, results in a product that is the identity matrix, $\mathbf{I}$.

- **Matrix subtraction** – the difference of two matrices, taken element by element.

- **Maximum of a set** – the largest numerical value of a set.

- **Mean** – the sum of the individual measurements, weighted by the relative frequency of its occurrence.

- **Measure of central tendency** – quantitative description of the way in which data is grouped toward the center of its range.

- **Median** – when a set containing N values is arranged in order numerically, the value dividing the set into two components, half of which are greater, half of which are less. If N is odd, it is the $\frac{(N+1)}{2}$ value. If N is even, it is the average of the $\frac{N}{2}$ and the $\frac{N}{2} + 1$ values.

- **Method of corners** – the process of optimizing a function by examining the values of the objective function at each corner of the feasible set.

- **Method of least squares** – (same as least-squares method)

- **Method of substitution** – reducing the number of equations by solving one equation for a single variable, then substituting this value into the remaining equations.

- **Minimum of a set** – the smallest numerical value of a set.

- **Mode** – the most frequently occurring numerical value.

- **Mortgage** – borrowing money from a lender, using the commodity as equity.

- **Multiplication principle** – counting the number of possible ways in which a multi-step task can be accomplished by taking the product of the ways in which each step can be accomplished.

- **Non-singular matrix** – a matrix that possesses an inverse.

- **Normal distribution** – a continuous probability distribution with the probability density function

$$P(x) = Ae^{-\frac{1}{2}(x-\mu)^2/\sigma^2}$$

- **Objective function** – a function that is maximized (such as profit) or minimized (such as lost inventory).

- **Order** – taking into account the sequence in which an event occurs.

- **Outcome** – a single observation, the result of a single experiment.

- **Outlier** – a value, or coordinate, which is far away from the rest of the values. In particular, a point which is greater than 1.5 times the interquartile range below the first quartile or above the third quartile. Alternately, a value which is more than three times the standard deviation away from the mean.

- **Over-determined system of equations** – a system of equations that has more equations than dependent variables.

- **Parallel** – in two dimensions, two lines that have the same slope and do not intersect.

- **Permutation** – the number of ways an event can occur, taking into account order.

- **Perpendicular** – in two dimensions, two lines that meet at right (90 degree) angles. Two lines with the property that their slopes satisfy $m_1 \times m_2 = -1$ where m is the slope of a line.

- **Pivoting** – performing a series of elementary row operations in such a way that all elements above and below a particular element vanish.

- **Point-slope form of a line** – an equation of the form $y - y_1 = m(x - x_1)$

- **Polygon** – a geometric figure whose sides are straight lines.

- **Polyhedra** – a geometric figure whose sides are planes lines.

- **Principal** – the amount of money, at the beginning of each interest period, for which interest is calculated.

- **Probability distribution** – the range of probabilities assigned to events in a sample space.

- **Proper subset** – a set which is a subset of another set, but not equal to it.

- **Quadratic regression** – determining the quadratic function which gives the most probable predicted value for a given y, based on a set of (x, y) values.

- **Quantitative description of data** – studying relationships between sets of data; specifically, relationships that can be reduced to numerical statements.

- **Quartiles** –
  - First quartile – the value below which 25% of the data lies.
  - Second quartile – the value above (and below) which 50% of the data lies.
  - Third quartile – the value above which 25% of the data lies.

- **Random variable** – a function that assigns a real number to an event.

- **Range** – the set of values a function takes on.

- **Range of a data set** – difference between the maximum of a set and the minimum if the set of data consists of real numbers. If the data consists of integers, the range of the data is the maximum value – minimum value + 1.

- **Rate of depreciation** – the slope of the linear depreciation curve.

- **Real number** – a number that is an integer, or a rational (fraction), or the limit of a sequence of rationals.

- **Regression line** – the linear function which generates the most probable predicted value for a given y, based on a set of (x, y) values.

- **Relative frequency** – the number of times a given outcome *actually occurs*.

- **Relative maximum** – a function value that is greater than or equal to all neighboring values in the range.

- **Relative minimum** – a function value which is less than or equal to all neighboring values in the range.

- **Rise** – the change in the ordinate (y-axis) of a linear function.

- **Root mean square** – the square root of the average of the squares.

- **Roster notation** – description of a set by means of a direct listing of elements.

- **Row matrix** – a matrix consisting of exactly one row.

- **Row operations** – one of the following three operations:

  o Interchanging two rows of a matrix

  o Multiplying a row by a scalar

  o Adding two rows of a matrix together

- **Run** – the change in the abscissa (x-axis) of a linear function.

- **Sample point** – a single member of a sample space, associated with a single outcome.

- **Sample space** – the universal set consisting of all possible outcomes of an experiment.

- **Scalar multiplication** – the process of multiplying a matrix by a scalar (real) number.

- **Scatter plot** – a graph consisting of discrete points whose location is given by a set of ordered pairs.

- **Set** – an unordered group or collection of objects.

- **Set-builder notation** – a description of a set by means of a rule for membership.

- **Set membership** – a requirement that an object must satisfy in order to belong to a set.

- **Simple frequency distribution** – a graph of the number of times each individual value of a set occurs.

- **Simple interest** – a process in which interest is calculated by multiplying the principal by the interest rate over the interest period.

- **Simplex algorithm** – an algorithm that optimizes an objective function by examining its values at a particular sequence of vertices of the feasible set.

- **Singular matrix** – a matrix that does not possess an inverse.

- **Slack variables** – additional variables that are introduced in order to make an inequality constraint into an equality constraint.

- **Slope** – the ratio of the change in the y-value to the change in the x-value.

- **Slope-intercept form of a line** – an equation of the form $y = mx + b$.

- **Spread of a distribution** – a quantitative measure of the range of data.

- **Square matrix** – a rectangular array of values consisting of an equal number of rows and columns.

- **Standard deviation** – the mean distance of data away from the mean.

- **Standard form of a line** – a linear equation of the form $Ax + By + C = 0$.

- **Standard normal distribution** – a normal probability distribution with a mean of zero and a standard deviation of one.

- **Standard normal variable** – a random variable that can be described by the standard normal distribution.

- **Subset** – a set, each of whose members is also in another set.

- **Summation notation** – the sum of the first N values of an indexed variable, x_i, is given by

$$x_1 + x_2 + \ldots + x_N = \sum_{i-1}^{i=N} x_i.$$

- **Supply curve** – the relationship between the amount of a commodity available for meeting a demand versus the price.

- **System of linear equations** – a set of more than one linear equation.

- **Theoretical probability** – the frequency that an event should occur in the limit of an infinite number of trials. Based on laws of probability, not observation.

- **Transpose of a matrix** – the matrix obtained from a given matrix by interchanging rows and columns.

- **Traveling salesman problem** – an optimization problem that requires an individual to visit a certain number of locations in such a way that the total distance is minimized.

- **Tree diagram** – a representation of all possible outcomes of an experiment by means of a graph.

- **Undefined slope** – a line that has a run that vanishes. A vertical line.

- **Uniform probability distribution** – a probability distribution for which all the probabilities are the same.

- **Uniform sample space** - a sample space in which all events have equal probability

- **Uniformly distributed random variable** – a function that assigns a real number to events arising from a uniform probability distribution

- **Union** – the set of all elements belonging to one set *or* another *or* both.

- **Universal set** – the set containing all possible values of interest.

- **Value** – an assigned or calculated quantity.

- **Variability of a distribution** – how far away from the mean the data gets.

- **Variance** - the square of the standard deviation.

- **Venn diagram** – representation of set membership by inclusion in one or more overlapping circles.

- **Vertical line** –a line with undefined slope. A set of points with constant abscissa (x-values).

- **Zero matrix** – a matrix in which all elements vanish.

Answers to Odd-Numbered Exercises

1.1 $y = -1$

1.3 $x = 2$

1.5 $a = 5$

1.7 $(3, 0)$

1.9 $S(x) = 0.00125x + 0.75$

2.1 $y = (223/124)x + (165/62) \approx 1.7984x + 2.6613$

2.3 $y = (326/35)x + (191/15) \approx 9.3143x + 12.7333$

2.5 Sales are unlikely to increase steadily year to year, so a linear model would not be appropriate to estimate next year's sales.

2.7 Approximately 26.86 hours

2.9 Yes

3.1 A system of two equations and two unknowns can have exactly one solution when the two lines intersect at exactly one point. The system can have no solution when the two lines are parallel. The last possibility is for the two lines to be the same, in which case there are infinitely many solutions corresponding to every point on the line.

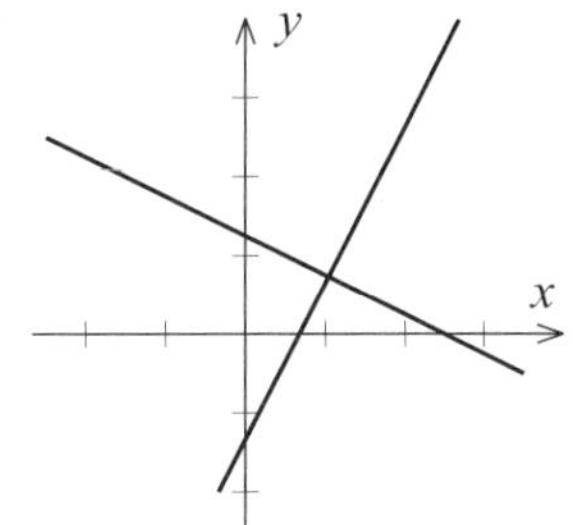

One Solution

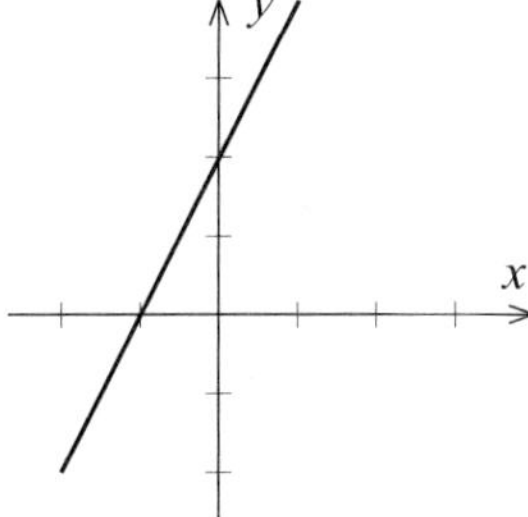

No Solution

Infinite Solutions

3. Solution is parametric, $(x, -3 - 1.5x)$ or $(-2 - (2/3)y, y)$

3.5 $a = 3.5, b = 5, c = 2, d = -8$

3.7 **a.** True **b.** False **c.** True **d.** True

3.9 $x = 2, y = -1, z = 0$

4.1

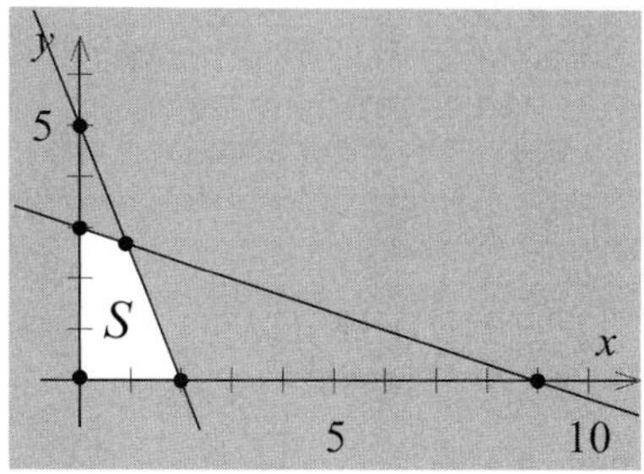

4.3   Maximum value of $P = 6$, minimum value of $P = -6$.

4.5   Make 0 rocking chairs and 16 lounging chairs.

4.7   Maximum value of $P = (150/7) \approx 21.43$.

4.9

$$\begin{bmatrix} 2 & -1 & \langle 3 \rangle & 1 & 0 & 0 & 0 & 24 \\ 1 & 5 & 0 & 0 & 1 & 0 & 0 & 15 \\ 3 & 0 & 2 & 0 & 0 & 1 & 0 & 18 \\ \hline -1 & -2 & -4 & 0 & 0 & 0 & 1 & 0 \end{bmatrix}$$

5.1   10

5.3   d

5.5

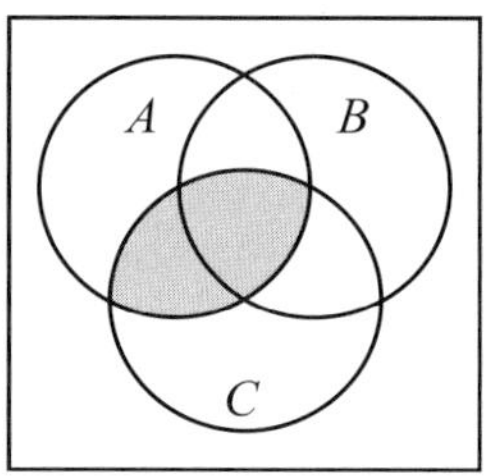

5.7

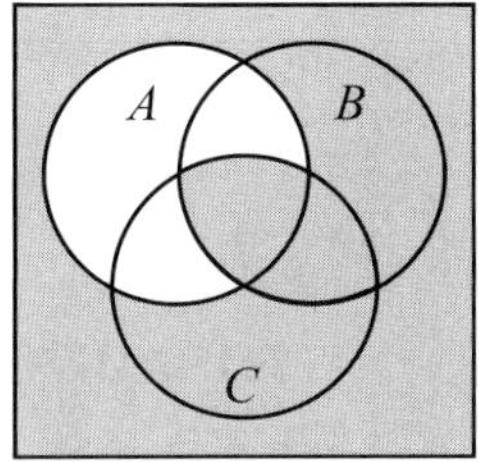

5.9   30

6.1   1,620
6.3 $2 \times 9! \times 9! \approx 2.6336 \times 10^{11}$
6.5   14
6.7   16
6.9 $27/646 \approx 0.0418$

7.1   2/5
7.3   6/91
7.5 $0.325 = 13/40$
7.7   1/3
7.9 $0.165 = 33/200$

8.1   52
8.3 $19/3 \approx 6.3333$
8.5   Mean
8.7   Histogram E: Mode = 6; Histogram F: Mode = 5, 25
8.9   12

9.1   0.2344
9.3   All positive integers
9.5   0.7333
9.7   d
9.9   Left endpoint: 8.5; Right endpoint: 200.5

10.1 $140.35
10.3 $145.37
10.5 $637.91
10.7 $1,168.06
10.9 $84,375.56

Answers to Sample Quizzes

1.1 $y = -2$

1.2 $x = 3$

1.3 $x = -1$

1.4 $x = -2$

1.5 $a = -1$

1.6 $y = -x/m + 1$

1.7 $(-2, -13)$

1.8 $D(x) = p = (-x + 38)/9$

1.9 $S(x) = p = 5x/3 + 20$

1.10 So, the equilibrium price will be 4.00 Utopian bucks, and 2,000,000 liters of fuel (x was given in millions of liters) will be sold at this price.

2.1 A scatter plot of the data shows that the data follows a general upward trend. Linear models are often useful because of their simplicity and so the least-squares line is a good starting place to model this data.

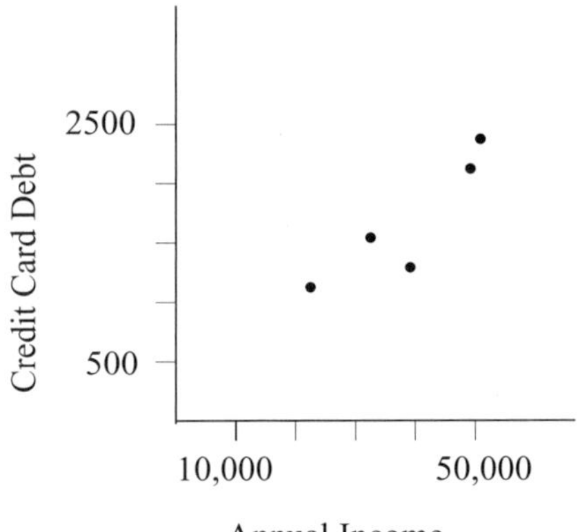

2.2 $1,760.70

2.3 $43,571.79

2.4 A correlation coefficient of .89 means that the data is reasonably well modeled with a linear function. Since the sign of the correlation coefficient is positive, as the x-value of our model increases, the y-value also increases. For this example, $r = .89$ says that as annual income increases, the credit card debt also increases.

2.5 The regression equation is $y = (114/35)\, x - 2 \approx 3.257\, x - 2$.

2.6 If each penny weighs 3 grams, then the weight of x pennies will be $3x$ and so $y = 3x$ would be the equation for a perfectly accurate scale.

2.7 For 12 pennies we have $x = 12$. Use this value of x in the least-squares equation, $y = 3.257 \times 12 - 2 = 37.084$ or a weight of about 37.1 grams.

2.8 The scale seems to be fairly accurate near the middle, but seems to weigh too low for small weights and slightly too much for heavier weights.

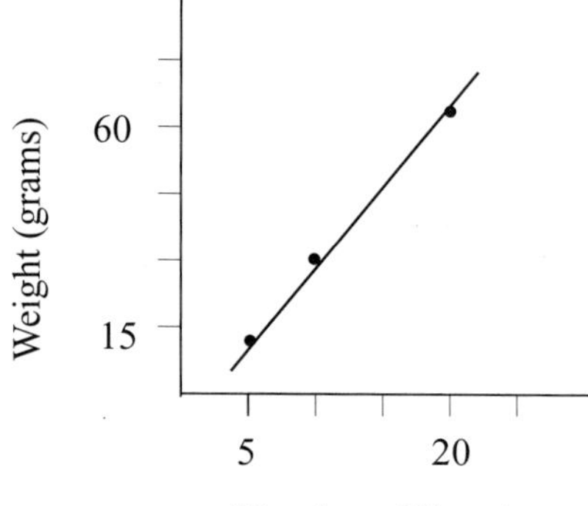

2.9 Data that falls exactly on a line with positive slope will have a correlation coefficient of 1 and data that falls exactly on a line with negative slope will have a correlation coefficient of -1. Therefore, if the correlation coefficient is close to either 1 or -1, the data is very linear and the least-squares line is likely a good model for the data.

2.10 A correlation coefficient that is close to zero means that the data is not linear and perhaps a non-linear model would be better.

3.1 **a.** C **b.** A **c.** B

3.2 No solution

3.3 One solution, (28/3, 4)

3.4 Solution is parametric, $(x, 4 + 2x)$ or $(-2 + 0.5y, y)$

3.5 $w = 8$, $x = 3$, $y = -11$, $z = -9$

3.6 **a.** 1×1 **b.** 5×5 **c.** 1×5 **d.** Does not exist
     e. Does not exist   **f.** Does not exist   **g.** Does not exist   **h.** 5×1

3.7 **a** and **b** are false statements.

3.8 **a** and **e** must be met.

3.9 $x = 0$, $y = -1$, $z = 2$

3.10 Solution is parametric, $(2 - 0.5z, -1 + 2z, z)$

4.1 The feasible region has 3 corners.

4.2

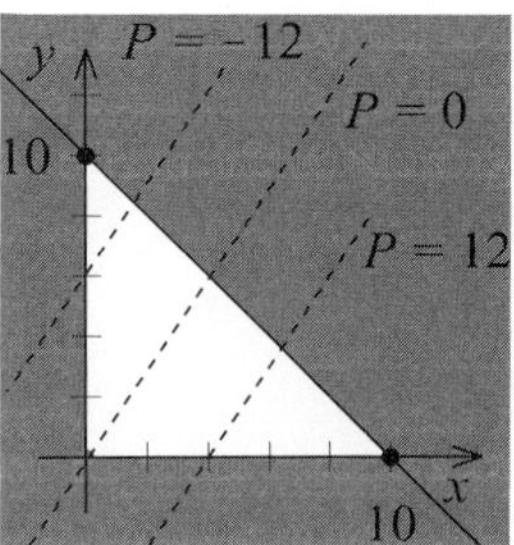

The objective function is increasing in the direction of increasing x.

4.3 The corners are at $(0, 0)$, $(10, 0)$, $(0, 10)$.

4.4 The objective function is maximized at $(10, 0)$.

4.5 The vertex for the maximum occurs at the intersection of the line $y = 0$ and the line $x + y = 10$.

4.6 The simplex tableau looks like

$$\begin{bmatrix} 3 & 1 & 2 & 1 & 0 & 0 & 0 & 9 \\ 2 & 3 & 1 & 0 & 1 & 0 & 0 & 8 \\ 1 & 2 & 3 & 0 & 0 & 1 & 0 & 7 \\ -20 & -12 & -18 & 0 & 0 & 0 & 1 & 0 \end{bmatrix}$$

4.7 There are 7 variables.

4.8 The first pivot element is in column 1, row 1. After the pivot the tableau looks like

$$\begin{bmatrix} 1 & 1/3 & 2/3 & 1/3 & 0 & 0 & 0 & 3 \\ 0 & 7/3 & -1/3 & -2/3 & 1 & 0 & 0 & 2 \\ 0 & 5/3 & 7/3 & -1/3 & 0 & 1 & 0 & 4 \\ 0 & -16/3 & -14/3 & 20/3 & 0 & 0 & 1 & 60 \end{bmatrix}$$

The value of the objective function is 60.

4.9   The second pivot element is in column 2, row 2. After the pivot the tableau should look like

$$\begin{bmatrix} 1 & 0 & 5/7 & 3/7 & -1/7 & 0 & 0 & 19/7 \\ 0 & 1 & -1/7 & -2/7 & 3/7 & 0 & 0 & 6/7 \\ 0 & 0 & 18/7 & 1/7 & -5/7 & 1 & 0 & 18/7 \\ \hline 0 & 0 & -38/7 & 36/7 & 16/7 & 0 & 1 & 452/7 \end{bmatrix}$$

The value of the objective function is $452/7 \approx 64.6$.

4.10   The final pivot element is in column 3, row 3. After the pivot the tableau should look like

$$\begin{bmatrix} 1 & 0 & 0 & 7/18 & 1/18 & -5/18 & 0 & 2 \\ 0 & 1 & 0 & -5/18 & 7/18 & 1/18 & 0 & 1 \\ 0 & 0 & 1 & 1/18 & -5/18 & 7/18 & 0 & 1 \\ \hline 0 & 0 & 0 & 49/9 & 7/9 & 19/9 & 1 & 70 \end{bmatrix}$$

The maximum value of the objective function is 70.

5.1   b and c

5.2 $n(A) = 3, n(A^c) = 7$

5.3 $n(A \cap B) = 4, n(A^c \cap B) = 6, n(A \cup B^c) = 17$

5.4 $n(A \cap B) = 20, n(A^c \cap B) = 15, n(A \cup B^c) = 85$

5.5   **a.** 50   **b.** 5   **c.** 3

5.6

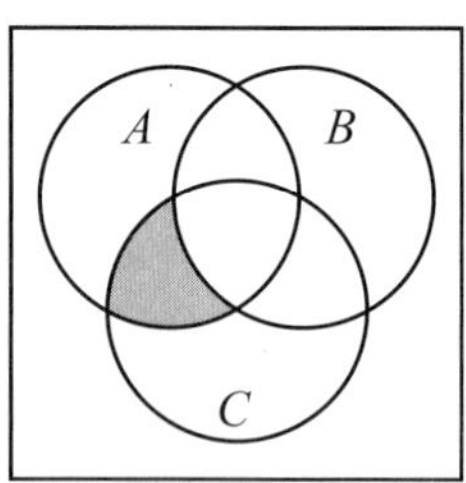

5.7

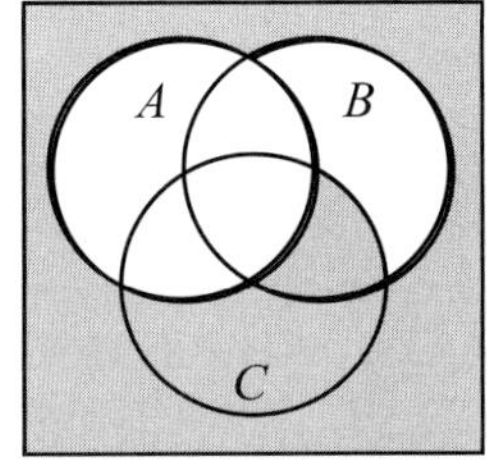

5.8

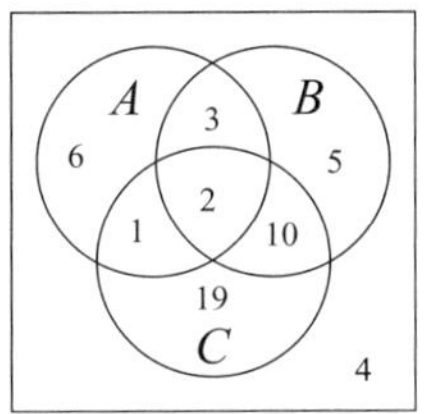

5.9   **a.** 30 **b.** 23 **c.** 47

5.10   **a.** 67**b.** 25**c.** 27

6.1   144

6.2   5,040

6.3   28,800

6.4   3,360

6.5   72

6.6   84

6.7   8

6.8 $5/11 \approx 0.4545$

6.9 $3/11 \approx 0.2727$

6.10 $7/18 \approx 0.3889$

7.1 $2/10 = 1/5$

7.2   0.7

7.3   27/35

7.4   3/11

7.5 $0.58 = 29/50$

7.6   4/51

7.7   2/3

7.8   8/17

7.9 $0.21 = 21/100$

7.10   27/79

8.1   98

8.2   5

8.3   20

8.4   7

8.5   Mean

8.6   4

8.7   Histogram P: Mode = 4; Histogram Q: Mode = 6, 8

8.8   Histogram P

8.9   2 blue worms

8.10 $98/39 \approx 2.5128$

9.1   0.2188

9.2   0.4962

9.3   All positive integers

9.4   No

9.5 $\mu = 32$, $\sigma = 2.5298$

9.6   Left endpoint = 60.5; Right endpoint = 65.5

9.7   0.3657

9.8   0.0082

9.9   0.1056

9.10 $X = 1, 2, 3, \ldots, 22$

10.1   \$1,540.14

10.2   \$1,577.63

10.3   \$1,577.73

10.4   4.3% compounded monthly

10.5   \$557.08

10.6   \$5,424.80

10.7   \$6,181.95

10.8   \$881.95

10.9   \$255,800.59

10.10   \$155,800.59

Index

A

B

C

D

E

N

O

P

Q

R